WITHDRAWN

The Pesticide Book

The Pesticide Book

GEORGE W. WARE

UNIVERSITY OF ARIZONA

W. H. FREEMAN AND COMPANY
San Francisco

Library of Congress Cataloging in Publication Data

Ware, George Whitaker, 1927–
  The pesticide book.

  Bibliography: **p.**
  Includes index.
    1. Pesticides. I. Title. [DNLM: 1. Pest
control—Handbooks. 2. Pesticides—Handbooks.
SB951 W268p]
SB951.W318   628.9   78-16220
ISBN 0-7167-0198-7

Printed in the United States of America

9 8 7 6 5 4 3

To my father and mother,
George and Dorothy

Contents

Preface

Pesticides, those "superchemicals" used to control pests around the home as well as in agriculture, have a prominent place in the day-to-day activities of our technologically advanced society. They can't be ignored, and they won't go away. Like it or not, we find ourselves, in the last quarter of the twentieth century, beginning what historians will refer to as "The Chemical Age." Time, population density, and technology have isolated us from the "good old days" (whenever they were), and we could no more return to the nonchemical, back-to-nature way of life than we could park our automobiles and walk everywhere we would normally drive.

Because pesticides are now a way of life and are essential ingredients in our affluent lives both here and abroad, it becomes necessary for every educated person, every conscientious citizen, to know something about these valuable chemicals. In agriculture, pesticides have become absolutely essential, as have the tractor, mechanical harvesters, and automatic milking machines. It would be utterly impossible to return to farming practices of as recently as 1960. There is simply not enough available, willing humanpower to plant, thin, cultivate, fertilize, weed, irrigate, and harvest our food crops as we did in 1960. For example, in 1960 one farm worker produced enough commodities to supply 25 others; in 1976 he or she produced enough for 56.

Although practically everyone seems to know something about pesticides, this familiarity is generally fragmentary, and thus understanding is frequently less than adequate. Often concepts are wrong. My experiences in talking to garden clubs, civic groups, and college seminars convinced me that everyone enjoyed learning even the most intricate details of pesticide chemistry if the information were presented along familiar paths. It seemed, then, that the subject matter of pesticides could be prepared in an easily understood, handbook form for persons from varied backgrounds. This book is the fruit of my efforts to accomplish that enormous task.

Its purpose is to present pesticides to laypeople in a simple, understandable way—to give an appreciation for these very important chemical tools and the fine state of this segment of the chemical arts. Discussions of individual compounds or groups are not restricted to bare definitions—sufficient information is given to convey something of their importance to our society.

This book is not just a catalog of select pesticides, since several well-constructed listings are available elsewhere. The individual pesticides used for illustrating certain points were not selected to the exclusion of others but were chosen, perhaps a bit arbitrarily, for such memorable characteristics as a notorious history, catchy name, or unique chemistry. The selection of individual pesticides in no way indicates an endorsement of those illustrated or a rejection of those not discussed.

While writing this fundamental book on pesticides and their important position in today's technology, I have attempted to adopt an everyday, factual, yet unemotional approach to a subject that is more often than not presented in a highly emotionally charged manner. Also, an effort has been made to avoid the use of technical and scientific terms prior to their introduction and discussion. In trying to avoid the natural tendency toward oversimplification, however, it was necessary to compromise between the too simple and the incomprehensible.

No book on this subject could be complete, because of the tremendous amount of material to be covered. This book is intended to present a comprehensive picture of pesticides as a subject and is not intended as an exhaustive study of any individual class of pesticides. It will be of interest not only to the householder but also to gardners, groundskeepers, landscape maintenance persons, structural pest control specialists, and to students enrolled in ecological and environmental studies courses.

A great many ideas and some data are presented in various sections of the book without direct citation of sources. After the glossary is a bibliography that lists the contributors whose work or writings were used.

During the writing of this book I received valuable advice, criticism, and assistance from many people. In particular I want to express my appreciation to Dr. Larry P. Pedigo of Iowa State University and Dr. Lena B. Brattsten of Cornell University for reviewing the manuscript. Special recognition is given to Mrs. Hazel Tinsley, who typed the manuscript more carefully than if it were her own.

Finally, to Doris, Cindy, Sam, and Julie, my wife and children, who sacrificed evenings and weekends of sharing time, thus enabling me to see this book to its completion, I owe a great debt of appreciation.

Tucson, Arizona   *George W. Ware*
May 1978

PART 1

BACKGROUND:
NAMES AND PERSONALITIES
OF PESTICIDES

ndrin   barban   monuron   naptalam   warfarin   zira

rdane   malathion   endosulfan   acrolein   naptalam   ph

none   sesamex   isodrin   schradan   paradichlorobe

nicotine   lindane   dieldrin   demeton   aldicarb   bina

non   heptachlor   ryania   butachlor   chlorazine   toxa

othall   fenac   benomyl   cyclohexamide   strychnine

anil   simazine   paraquat   captan   fluometuron   dichl

CHAPTER 1

Pesticides: Chemical Tools

Let's get our priorities in perspective....
We must feed ourselves and protect ourselves
against the health hazards of the world.
To do that, we must have agricultural chemicals.
Without them, the world population will starve.

Norman E. Borlaug,
1970 Nobel Peace Prize

Pesticides are chemical substances used to kill or control pests. To the grower or farmer, pests could include insects and mites that damage crops; weeds that compete with field crops for nutrients and moisture; aquatic plants that clog irrigation and drainage ditches; diseases of plants caused by fungi, bacteria, and viruses; nematodes, snails, and slugs; rodents that feed on grain, young plants, and the bark of fruit trees; and birds that eat their weight every day in young plant seedlings and grain from fields and feedlots as well as from storage.

To the homeowner, pests may include filthy, annoying, and disease-transmitting flies, mosquitoes, and cockroaches; moths that eat woolens; beetles that feed on leather goods and infest package foods; slugs, snails, aphids, mites, beetles, caterpillars, and bugs feeding on lawns, gardens, and ornamentals; termites that nibble away at wooden buildings; diseases that mar and destroy plants; algae growing on the walls or clouding the water of swimming pools; slimes and mildews that grow on shower curtains and stalls and under the rims of sinks; rats and mice that leave their fecal pellets scattered around in exchange for the food they eat; dogs that designate their territories by urinating on shrubs and favorite flowers; alley cats that yowl and screech at night; and annoying birds that defecate on window ledges, sidewalks, and statues of yesterday's forgotten heroes.

Pesticides are big business. The U.S. market is the world's largest, utilizing between 35 and 45 percent of the total, having reached

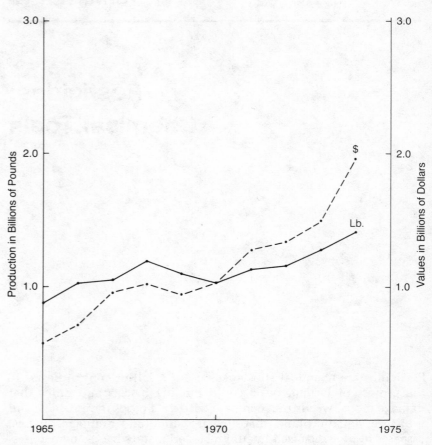

FIGURE 1
Production and sales of organic pesticides by the United States, 1965–1974.
(*Source:* D. L. Fowler and J. N. Mahan. "Pesticide Supply, Production, Use and Trade," *The Pesticide Review 1975,* July 1976, p. 2. U.S. Department of Agriculture, Agricultural Stabilization and Conservation Service, Washington, D.C.)

$1.82 billion in 1974, the value at the user level (Figure 1). This figure is expected to reach $2.3 billion by 1980. Although our own domestic market has grown substantially, the rest of the world is utilizing pesticides at an increasingly faster rate as more and more countries develop their economies.

The reader may be surprised to learn that the agricultural market consumes 92 percent of the pesticides sold in the United States. The remaining 8 percent or nonagricultural portion amounted to a $106 million business in 1973. This nonfarm market is composed of pest control operators, turf and sod producers, floral and shrub nurseries, railroads, highways, utility rights-of-way, and industrial plant sites. With the exception of pest control operators, these are primarily herbicide markets. There are no accurate figures on the household and lawn market, since the value of the pesticides sold to you, the homeowner, is a very small part of the total. However, a recent survey in Arizona indicated that homeowners used pesticides at about the same rate per acre of home and garden as did the agricultural growers from that productive state. With regard to value of total sales by classes, herbicides led with 58 percent, followed by insecticides (which includes fumigants and rodenticides), amounting to 35.5 percent, and fungicides, 6.8 percent (Fowler and Mahan, 1976).

Of the nonfarm market, insecticides dominate, with 46 percent of the total, followed by herbicides with a 41 percent share. Part of the explanation for insecticides exceeding herbicides, is their use by pest control operators who specialize in controlling termites and other urban pests, using insecticides, herbicides, and rodenticides, in that order.

Although most pesticides are synthetic, a few are produced naturally by plants. The U.S. Environmental Protection Agency (EPA) has more than 1,200 pesticides registered as of 1976. Of these 275 are herbicides, 400 are insecticides, 200 are fungicides and nematicides, 100 are rodenticides, and 225 are disinfectants. These are sold in the form of 30,000 products or formulations, and 5 pounds of pesticides are used each year to feed, clothe, and protect every man, woman, and child in the United States alone. Part of these 5 pounds are used at home, and that's what this book is about.

Pesticides have become extremely beneficial tools for the suburbanites, the homeowners. They depend on pesticides, perhaps, more than they realize: for algae control in the swimming pool, weed control in the lawn, flea powder for pets, sprays for controlling a myriad of garden and lawn insects and diseases, household sprays for ants and roaches, aerosols for flies and mosquitoes, soil and wood treatment for termite protection by professional exterminators, baits for the control of mice and rats, woolen treatment at the dry cleaners for clothes moth protection, and repellents to keep off biting flies, chiggers, and mosquitoes when camping or fishing.

IMPORTANCE OF PESTICIDES
AND THEIR DOLLAR VALUE

Down on the farm, however, pesticides have become essential tools. Just as the tractor, mechanical harvester, electric milker, and fertilizers are part of modern agricultural technology, so too, are pesticides.

These chemical tools are used as intentional additions to the environment in order to improve environmental quality for ourselves, our animals, and our plants. Pesticides are used in agriculture to increase the ratio of cost/benefit in favor of the grower and, ultimately, the consumer of food and fiber products—the public. Pesticides have contributed significantly to the increased productive capacity of U.S. farmers, each of whom produced food and fiber for 3 persons in 1776 and for 56 persons in 1976 (Figure 2).

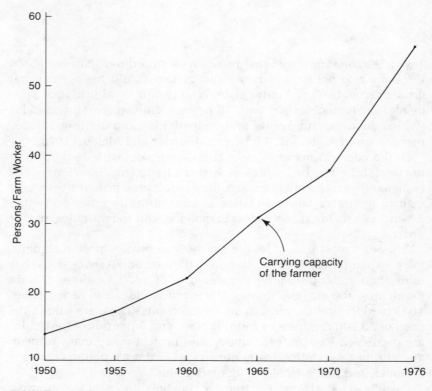

FIGURE 2
Persons supplied farm products per farmer, 1950–1976. (*Source:* Agricultural
Statistics, 1976. U.S. Department of Agriculture, Statistical Reporting Service. U.S.
Superintendent of Documents, U.S. Government Printing Office, Washington, D.C.,
p. 56.)

They have rapidly evolved as extremely important aids to world
agricultural production. It is estimated that insects, weeds, plant
diseases, and nematodes account for losses of up to $15 billion annu-
ally in the United States alone. The use of these pesticidal chemicals
in agriculture makes it possible to save approximately an overall
one-third of our crops. It is because of the economic implications
of such losses and savings that pesticides have assumed their
importance.

IMPORTANCE OF PESTS AND THEIR DAMAGE

The world's main source of food is plants. They compete with about
80,000 to 100,000 plant diseases caused by viruses, bacteria, mycoplasmalike organisms, rickettsias, fungi, algae, and parasitic seed
plants; 30,000 species of weeds the world over, with approximately
1,800 species causing serious economic losses; 3,000 species of
nematodes that attack crop plants, with more than 1,000 that cause
damage; and over 800,000 species of insects, of which 10,000 species
add to the devastating loss of crops throughout the world.

Approximately one-third of the world's food crops is destroyed by these pests during growth, harvesting, and storage. Losses are even higher in emerging countries: Latin America loses to pests approximately 40 percent of everything produced. Cocoa production in Ghana, the largest exporter in the world, has been trebled by the use of insecticides to control just one insect species. Pakistan sugar production was increased 33 percent through the use of insecticides. The Food and Agriculture Organization (FAO) has estimated that 50 percent of cotton production in developing countries would be destroyed without the use of insecticides.

Good examples of specific increases in yields resulting from use of insecticides in the United States are: cotton, 100 percent; corn, 25 percent; potatoes, 35 percent; onions, 140 percent; tobacco, 125 percent; beet seed, 180 percent; alfalfa seed, 160 percent; and milk production, 15 percent.

Equally important are the agricultural losses from weeds. They deprive crop plants of moisture and nutritive substances in the soil. They shade crop plants and hinder their normal growth. They contaminate harvested grain with seeds that may be poisonous to humans and animals. In some instances, complete loss of the crop results from disastrous competitive effects of weeds.

PESTS IN HISTORY

History contains innumerable examples of the mass destruction of crops by diseases and insects. In the period from 1845 to 1851, the potato famine in Ireland occurred, as a result of a massive infection of potatoes by a fungus, *Phytophthora infestans*, now commonly referred to as *late blight*. This resulted in the loss of about a million lives and the cultural invasion of America by the Irish. Surprisingly, the infected potatoes were edible and nutritious, but the superstitious population refused to use diseased tubers. In 1930, 30 percent of the U.S. wheat crop was lost to stem rust, the same disease that destroyed three million tons of wheat in western Canada in 1954.

Aside from the historical plant epidemics, let's examine some of our more recent problems, animal and human disease, caused by organisms that are carried by insects. In 1971, Venezuelan equine encephalitis appeared in southern Texas, moving in from Mexico. However, through a very concerted suppression effort, involving horse vaccination, a quarantine on horse movement, and extensive spraying for mosquito control, the reported cases were limited to 88 humans and 192 horses. With the other arthropod-borne encephalitides, there were an average of 205 human cases in the United States annually between 1964 and 1973.

As late as 1955, malaria infected more than 100 million persons throughout the world. The annual death rate from this debilitating disease has been reduced from 6 million in 1939 to 2.5 million today. Similar progress has been made in controlling other important tropi-

cal diseases, such as yellow fever, sleeping sickness, and Chagas'
disease, through the use of insecticides.

In the nineteenth century, the Panama Canal was abandoned by the
French because more than 30,000 (think of it!—30,000) of their
laborers died from yellow fever! Since the first recorded epidemic of
the Black Death, or bubonic plague, it is estimated that more than 65
million persons have died from this disease, which is transmitted by
the rat flea. The number of deaths resulting from all wars appears
paltry beside the toll taken by insect-borne diseases.

And currently there is the ever-lurking danger to humans from
such diseases as encephalitis, typhus, relapsing fever, sleeping sickness, elephantiasis, and many others, which are transmitted by insects or mites.

FOOD AND HUNGER

It is common knowledge that our current world food supply is
inadequate. As much as 56 percent of the world's population is
undernourished. And the situation is worse in undeveloped countries, where an estimated 79 percent of the inhabitants are undernourished.

The earth's population was estimated at 3.57 billion in 1970, and
is expected to reach 4.27 billion by 1980, 5.07 billion by 1990.
These numbers are not intended to evoke gloom about our ability to
support such a population but to suggest that there will be great
pressure to increase agricultural production, since this extra population must be fed and clothed.

NEED FOR PESTICIDES

When millions of humans are killed or disabled annually from
insect-borne diseases and world losses from insects, diseases, weeds,
and rats are estimated at $75 billion annually, it becomes obvious to
the reader that control of various harmful organisms is vital for the
future of agriculture, industry, and public health. Pesticides thus
become indispensable in feeding, clothing, and protecting the
world's population, which is predicted to double by the year 2000.

The Vocabulary of Pesticides

After comprehending the importance of pesticides and recognizing the need for knowledge of these essential but mysterious chemicals, let us pursue the game of learning what pesticides are all about. To one person, the word *pesticide* may suggest the insecticide DDT. To another it may conjure up the herbicide 2,4-D, and to still another the fungicide maneb. All are correct—however, only in part, for the uses and effects of these three materials are totally unrelated.

The term *pesticides* is an all-inclusive but nondescript word meaning "killer of pests." The various generic words ending in *-icide* (from the Latin *-cida*, "to kill") are classes of pesticides, such as insecticides and herbicides. Table 1 lists the various pesticides and other classes of chemical compounds not commonly considered pesticides. These others, however, are included among the pesticides as defined by federal and state laws.

Pesticides are legally classed as "economic poisons" in most state and federal laws and are defined as "any substance used for controlling, preventing, destroying, repelling, or mitigating any pest." Should you ever pursue the subject of pesticides from a legal viewpoint, they would be discussed as economic poisons.

This is the end of the vocabulary section, which includes all of the commonly used pesticides, including groups of chemicals that do not actually kill pests. However, because they fit rather practically as well as legally into this umbrella word, *pesticides*, they are included.

If you have examined the table in detail with particular attention to the derivation of pesticide classes, your vocabulary should be somewhat expanded. Certainly your vocabulary should be more precise. and you should be able to distinguish between *pesticide* and *herbicide, insecticide* and *acaricide,* or defoliant and *growth regulator.*

TABLE 1

A list of pesticide classes, their use and derivation.

Pesticide class	Function	Root-word derivation
Insecticide	Controls insects	L.[a] *insectum,* "insecre, to cut or divide into segments"
Herbicide	Kills weeds	L. *herba,* "an annual plant"
Fungicide	Kills fungi	L. *mushroom,* Gr.[b] "spongos"
Nematicide	Kills nematodes	L. *nematoda* (Gr. *nema,* "thread")
Rodenticide	Kills rodents	L. *rodere,* "to gnaw"
Bactericide	Kills bacteria	L. *bacterium* (Gr. *baktron,* "a staff")
Acaricide	Kills mites	Gr. *akari,* "mite or tick"
Algicide	Kills algae	L. *alga,* "seaweed"
Miticide	Kills mites	Synonymous with *Acaricide*
Molluscicide	Kills snails and slugs (may include oysters, clams, mussels)	L. *molluscus,* "soft- or thin-shelled"
Avicide	Controls or repels birds	L. *avis,* "bird"
Slimicide	Controls slimes	Anglo-Saxon *slim*
Piscicide	Controls fish	L. *piscis,* "a fish"
Ovicide	Destroys eggs	L. *ovum,* "egg"

Chemicals classed as pesticides not bearing the -icide suffix

Disinfectants	Destroy or inactivate harmful micro-organisms	
Growth regulators	Stimulate or retard plant growth	
Defoliants	Remove leaves	
Desiccants	Speed drying of plants	
Repellents	Repel insects, mites and ticks, or pest vertebrates (dogs, birds)	
Attractants	Attract insects	
Chemosterilants	Sterilize insects	

[a] Latin origin.
[b] Greek origin.

CHAPTER 3

> I am a little world made cunningly
> of elements, and an angelic sprite.
>
> John Donne

The Chemistry of Pesticides

To better understand some of the chemistry of pesticides, it will be necessary to learn to identify the different elements with which pesticides are made. If you have had a course or two in chemistry, so much the better; if not, you are in for a revelation and a treat.

Everything about us, including the earth, is composed of chemical elements. These include such familiar elements as oxygen, nitrogen, iron, sulfur, and hydrogen. The smallest part or unit of an element is the atom. An element is pure in that all of its atoms are alike. It is not possible to change an element chemically.

A compound is a combination of two or more different elements. Water, salt, and DDT are examples of compounds. When different atoms combine or join, they form a molecule, the smallest part or unit of a compound. And because a compound is a combination of different elements, chemical changes can be made in it by changing the combinations of elements. Elements have a certain way of combining with other elements, and only certain elements will combine. Conversely, certain elements will not combine.

Combinations of two or more elements bound together by chemical bonds are chemical compounds. The most familiar compound is water, a mixture of hydrogen and oxygen, bound together by chemical bonds, which everyone recognizes as H_2O. Proper chemical terminology notes that H_2O is composed of two atoms of hydrogen bonded to one atom of oxygen to form one molecule of water.

To aid in the writing of chemistry, a chemical shorthand has been developed by chemists. Each element was given a symbol, which makes both the writing and the setting of type for printed matter simpler and easier to read. These symbols are frequently the first letter alone or the first letter plus another letter in its name.

Table 2 shows some of the elements or atomic names and the chemical symbols used when writing or illustrating molecules. The table has been specially prepared for this book and is incomplete, since it contains only those elements used in the making or synthesis of pesticides.

TABLE 2

Chemical elements from which pesticides are made.[a]

Atomic or element name	Symbol	Symbol derivation[b]
Arsenic	As	
Boron	B	
Bromine	Br	
Cadmium	Cd	
Carbon	C	
Chlorine	Cl	
Copper	Cu	Cuprum
Fluorine	F	
Hydrogen	H	
Iron	Fe	Ferrum
Lead	Pb	Plumbum
Magnesium	Mg	
Manganese	Mn	
Mercury	Hg	Hydrargyrum
Nitrogen	N	
Oxygen	O	
Phosphorus	P	
Sodium	Na	Natrium
Sulfur	S	
Tin	Sn	Stannum
Zinc	Zn	

[a] This list is incomplete and includes only those elements that appear in pesticides.
[b] Some atomic symbols are based on the Latin names.

As you become more familiar with pesticides, you will observe that only 21 of the more than 105 chemical elements are used as building blocks in their construction. This list can be further reduced by considering those elements used most frequently in pesticides: carbon, hydrogen, oxygen, nitrogen, phosphorus, chlorine, and sulfur. Some, however, may include the metallic and semimetallic elements, iron, copper, mercury, zinc, arsenic, and others. And because only about one-half the elements appearing in the table make up the bulk of the most frequently used pesticides, it becomes a simple matter to learn their symbols.

For practical generalizations, essentially all of the pesticides are organic compounds, that is, they contain carbon in their molecules. Only a few contain no carbon and are thus identified as inorganic compounds.

CHEMICAL FORMULAS

A chemical formula is the printed description of one molecule of a chemical compound. There are several types of chemical formulas that must be distinguished from each other, since all are used in this book.

The molecular formula uses the symbols of the elements to indicate the number and kind of atoms in a molecule of the compound, e.g., H_2O for water and C_6H_6 for benzene. The structural formula is written out, using symbols to indicate the way in which the atoms are located relative to one another in the molecule. For example, structural formulas for water and benzene can be presented as illustrated:

WATER

H—O—H

BENZENE

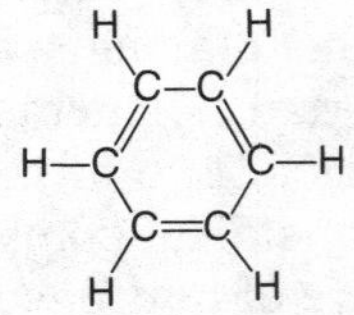

This six-carbon ring, with its six hydrogens as presented, is benzene and, for ease of representation, is usually indicated as a hexagon with double bonds. When other groups have replaced one or more of the hydrogens, the ring is referred to as the *phenyl radical* rather than as the *benzene radical*. For example, turn to the section on DDT (p. 29). Note that the old chemical name of DDT contains the words *di-phenyl,* which refers to the two benzene rings with other groups attached.

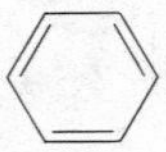

The benzene or phenyl ring is found quite frequently in pesticide structures, and, because it is an awkward structure for printers to set from type, simpler ways of presentation have been devised and are commonly used in pesticide literature. Two of these have already been shown. The four remaining designs are: (1) the hexagon containing a broken circle or oval; (2) similar but with an unbroken circle or oval; (3) the hexagon with the three double bonds extending to its sides; and, finally, (4) the Greek letter *phi*, Φ.

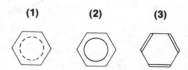

(1) **(2)** **(3)**

The actual shapes of organic molecules are not shown by structural formulas, but it is not essential that the shape and spatial design be known to become familiar with these structures. The structural formulas are shown for most of the pesticides discussed in this book.

Stereo or three-dimensional formulas are printed so that the reader can visualize the depth and spatial conformation of certain molecules. Printing the structural formula does not make the molecule flat, since the formula is a two-dimensional representation of a three-dimensional object.

The structural and stereo formulas for methane and norbornene are illustrated as examples:

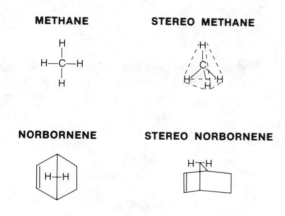

CHAPTER 4

What's in a name? That which we call a rose
By any other name would smell as sweet.
William Shakespeare, *Romeo and Juliet*

The Naming of Pesticides

A general knowledge of pesticides involves, among other things, learning not only about their structures, but their names or nomenclature. For example, let us look at chlordane, a commonly known household insecticide:

(1)
CHLORDANE

(2)
(Octa-Klor®)

(3)

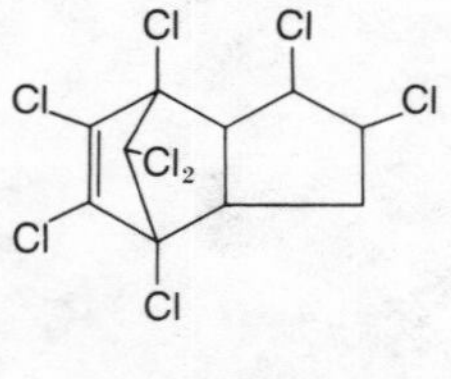

(4)
1,2,4,5,6,7,8,8-octachloro-3*a*,4,7,7*a*-tetrahydro-4,7-methanoindane

(5)

$C_{10}H_6Cl_8$

At the top is the name, chlordane (1), the common name for the compound. Common names are selected officially by the appropriate professional scientific society and approved by the American National Standards Institute (formerly United States of America Standards Institute) and the International Organization for Standardization. Common names of insecticides are selected by the Entomological Society of America; herbicides by the Weed Science

Society of America; and fungicides by the American Phytopathological Society. The proprietary name (2), trade name, or brand name for the pesticide is given by the manufacturer or by the formulator. It is not uncommon to find several brand or trademark names given to a particular pesticide in various formulations by their formulators. To illustrate, Octa-Klor® is (or was) also known as Octachlor, Velsicol 1068, and 1068. The latter was a code number assigned to the compound by the basic manufacturer when it was first synthesized in the laboratory.

Common names are assigned to avoid the confusion resulting from the use of several trade names, as just illustrated. The structural formula (3), as mentioned earlier, is the printed picture of the pesticide molecule. The long chemical name (4) beneath the structural formula is just that, the chemical name. It is usually presented according to the principles of nomenclature used in *Chemical Abstracts,* a scientific abstracting journal that is generally accepted as the world's standard for chemical names. And finally, (5), when used, is the molecular or empirical formula that indicates the various numbers of atoms for comparative purposes. Beyond this, the molecular formula will be illustrated only when the structural formula is not known, which is rare.

CHAPTER **5**

The Formulations of Pesticides

After a pesticide is manufactured in its relatively pure form—the technical grade material, whether herbicide, insecticide, fungicide, or other classification—the next step is formulation. It is processed into a usable form for direct application or for dilution followed by application. The formulation is the final physical condition in which the pesticide is sold for use. The technical grade material may be formulated by its basic manufacturer or sold to a formulator. The formulated pesticide will be sold under the formulator's brand name, or it may be custom-formulated for another firm.

Formulation is the processing of a pesticidal compound by any method that will improve its properties of storage, handling, application, effectiveness, or safety. The term formulation is usually reserved for commercial preparation prior to actual use and does not include the final dilution in application equipment.

The real test for a pesticide is acceptance by the user. And, to be accepted for use by the homeowner, the grower, or commercial applicator, a pesticide must be effective, safe, and easy to apply, but not necessarily economical, especially from the homeowner's viewpoint. The urbanite commonly pays 10 to 30 times the price that a grower may pay for a given weight of a particular pesticide, depending to a great extent on the formulation. For instance, the most expensive form of insecticide is the pressurized aerosol formulation.

Pesticides, then, are formulated into many usable forms for satisfactory storage, for effective application, for safety to the applicator and the environment, for ease of application with readily available equipment, and for economy. This is not always simply accomplished, due to the chemical and physical characteristics of the technical grade pesticide. For example, some materials in their "raw" or technical condition are liquids, others solids; some are stable to air and sunlight, whereas others are not; some are volatile, others not; some are water soluble, some oil soluble, and others may be insoluble in either water or oil. These characteristics pose problems to the formulator, since the final formulated product must meet the standards of acceptability by the user.

More than 98 percent of all pesticides used in the United States in 1976 are manufactured in the formulations appearing in the simplified classification presented in Table 3. Familiarity with the

TABLE 3
Common formulations of pesticides.[a]

1. Sprays (insecticides, herbicides, fungicides)
   a. Emulsible concentrates
   b. Water-miscible liquids
   c. Wettable powders
   d. Water-soluble powders, e.g., prepackaged, tank drop-ins
   e. Oil solutions, e.g., house and garden ready-to-use sprays
   f. Soluble pellets for water-hose attachments
   g. Flowable or sprayable suspensions
   h. Ultralow-volume concentrates (agricultural and forestry use only)
   i. Fogging concentrates, e.g., public health mosquito and fly abatement foggers

2. Dusts (insecticides, fungicides)
   a. Undiluted toxic agent
   b. Toxic agent with active diluent, e.g., sulfur
   c. Toxic agents with inert diluent, e.g., home garden insecticide-fungicide combination in pyrophyllite carrier
   d. Aerosol "dust," e.g., silica in aerosol form

3. Aerosols (insecticides, disinfectants, or "germicides")
   a. Pushbutton
   b. Total release

4. Granulars (insecticides, herbicides, algicides)

5. Fumigants (insecticides, nematicides)
   a. Stored products and space treatment, e.g., liquids, gases, and moth crystals
   b. Soil treatment liquids that vaporize

6. Impregnates (insecticides, fungicides, herbicides)
   a. Plastic strips containing a volatile insecticide, e.g., pet collars
   b. Shelf papers, strips, cords containing a volatile or contact insecticide
   c. Mothproofing agents for woolens (insecticides)
   d. Wood preservatives (fungicides, insecticides)
   e. Wax bars (herbicides)

7. Fertilizer combinations with herbicides, insecticides, or fungicides

8. Baits (insecticides, molluscicides, rodenticides)
   a. Insecticides, e.g., ants, roaches, wasps, crickets, grasshoppers, fruit flies
   b. Molluscicides, e.g., slugs, snails
   c. Rodenticides, e.g., mice, rats, gophers

9. Slow-release insecticides
   a. Encapsulated materials for agriculture and mosquito abatement
   b. Paint-on lacquers for pest control operators and homeowners
   c. Adhesive tapes for pest control operators and homeowners

10. Insect repellents
    a. Aerosols
    b. Rub-ons (liquids, cloths, and "sticks")
    c. Vapor-producing candles, torch fuels, smoldering wicks

11. Insect attractants
    a. Food (Japanese beetle traps)
    b. Sex lures (pheromones for agricultural and forest pests)

12. Animal systemics (insecticides, parasiticides)
    a. Oral (premeasured capsules or liquids)
    b. Dermal (pour-on or sprays)
    c. Feed-additive

13. Animal dressings (insecticides)

[a] This list is incomplete, containing only the more common formulations.

more important formulations is essential to the well-informed urbanite. We will now examine the major formulations used for the home,
lawn, and garden, as well as those employed in structural pest control and agriculture.

SPRAYS

Emulsible Concentrates

Formulation trends shift with time and need. Traditionally, pesticides have been applied as water sprays, water suspensions, oil
sprays, dusts, and granules. Spray formulations are prepared for insecticides, herbicides, miticides, fungicides, algicides, growth regulators, defoliants, and desiccants. Consequently, more than 75
percent of all pesticides are applied as sprays. The bulk of these are
currently applied as water emulsions made from emulsible concentrates, sometimes abbreviated as EC.

Emulsible concentrates, synonymous with emulsifiable concentrates, are concentrated oil solutions of the technical grade material
with enough emulsifier added to make the concentrate mix readily
with water for spraying. The emulsifier is a detergentlike material
that makes possible the suspension of microscopically small oil
droplets in water to form an emulsion.

When an emulsible concentrate is added to water, the emulsifer
causes the oil to disperse immediately and uniformly throughout the
water, if agitated, giving it an opaque or milky appearance. This
oil-in-water suspension is a normal emulsion. There are a few rare
formulations of invert emulsions, which are water-in-oil suspensions, and are opaque in the concentrated forms, resembling salad
dressing or face cream. These are employed almost exclusively as
herbicide formulations. The thickened sprays result in very little
drift and can be applied in sensitive situations.

Emulsible concentrates, if properly formulated, should remain
suspended without further agitation for several days after dilution
with water. A pesticide concentrate that has been held over from last
year can be easily tested for its emulsible quality by adding 1 ounce
to 1 quart of water and allowing the mixture to stand after shaking.
The material should remain uniformly suspended for at least 24
hours with no precipitate. If a precipitate does form, the same condition may occur in the spray tank, resulting in clogged nozzles and
uneven application. In the home situation, this can be remedied by
adding 2 tablespoons of liquid dishwashing detergent to each pint of
concentrate and mixing thoroughly. In an agricultural or other cir-
cumstance, where several gallons of costly pesticide are involved,
additional emulsifier can be obtained from the formulator. This
should be added to the concentrate at the rate of 1.0 to 1.5 pounds for
each gallon of outdated material. The bulk of pesticides available to
the homeowner are formulated as emulsible concentrates and generally have a shelf life of about 3 years.

Water-Miscible Liquids

Water-miscible liquids are exactly that, water mixable. The technical grade material may be initially water miscible, or it may be alcohol miscible and formulated with an alcohol to become water miscible. These formulations resemble the emulsible concentrates in viscosity and color, but do not become milky when diluted with water. Few of the home and garden pesticides are sold as water miscibles, since few of the pesticides that are safe for home use have these physical characteristics.

Wettable Powders

Wettable powders, abbreviated as WP, are essentially concentrated dusts containing a wetting agent to facilitate the mixing of the powder with water before spraying. The technical material is added to the inert diluent, in this case a finely ground talc or clay, in addition to a wetting agent, similar to a dry soap or detergent, and mixed thoroughly in a ball mill. Without the wetting agent, the powder would float when added to water, and the two would be almost impossible to mix. Because wettable powders usually contain from 50 to 75 percent clay or talc, they sink rather quickly to the bottom of spray tanks unless the spray mix is agitated constantly. Many of the insecticides sold for garden use are in the form of wettable powders because there is very little chance that this formulation can burn foliage, even at high concentrations. This is not true for emulsible concentrates, since the original carrier is usually an aromatic solvent, which in relatively moderate concentrations can cause foliage burning at high temperatures.

Water-Soluble Powders

Water-soluble powders (SP) are properly titled and self-explanatory. Here, the technical grade material is a finely ground water-soluble solid and contains nothing else to assist its solution in water. It is simply added to the spray tank, where it dissolves immediately. Unlike the wettable powders and flowables, these formulations do not require constant agitation; they are true solutions and form no precipitate.

Oil Solutions

In their commonest form, oil solutions are the ready-to-use household and garden insecticide sprays sold in an array of bottles, cans, and plastic containers, all usually equipped with a handy spray atomizer. Not to be confused with aerosols, these sprays are intended to be used directly on pests or where they frequent. Oil solutions may be used as roadside weed sprays, for marshes and standing pools to control mosquito larvae, in fogging machines for mosquito and fly abatement programs, or for household insect sprays purchased in supermarkets. Commercially they may be sold as oil concentrates of the pesticide to be diluted with kerosene or diesel fuel before application or in the dilute, ready-to-use form. In either case, the compound is dissolved in oil and is applied as an oil spray; it contains no dust diluent, emulsifier, or wetting agent.

Soluble Pellets

Despite their seeming convenience and ease of handling with a water hose, soluble pellets are not very effective. They are sold in kits, including the water-hose attachment, fertilizer, fungicide, insecticide, and even a car-wax pellet. The actual amount of active ingredients is very small, and even distribution with a watering hose is unlikely.

Flowable or Sprayable Suspensions

Flowable or sprayable suspensions (F or S) exemplify an ingenious solution to a formulation problem. Earlier it was stated that some pesticides are soluble in neither oil nor water. They are soluble in one of the exotic solvents, however, which makes the formulation quite expensive and may price it out of the marketing competition. To handle the problem, the technical material is blended with one of the dust diluents and a small quantity of water in a mixing mill, leaving the pesticide-diluent mixture finely ground but wet. This "pudding" mixes well with water and can be sprayed but has the same tank-settling characteristic as the wettable powders.

Ultralow-Volume Concentrates

Ultralow-volume concentrates (ULV) are available only for commercial use in the control of public health, agricultural, and forest pests. They are usually the technical product in its original liquid form or, if solid, the original product dissolved in a minimum of solvent. They are usually applied without further dilution, by special aerial or ground spray equipment that limits the volume from one-half pint to a maximum of one-half gallon per acre, as an extremely fine spray. The ULV formulations are used where good results can be obtained while economizing through the elimination of the normally high spray volumes, varying from 3 to 10 gallons per acre. This technique has proved extremely useful where insect control is desired over vast areas.

Fogging Concentrates

Fogging concentrates are the formulations sold strictly for public health use in the control of nuisance or disease vectors, such as flies and mosquitoes, and to pest control operators. Fogging machines generate droplets whose diameters are usually less than 10 microns but greater than 1 micron. They are of two types. The thermal fogging device utilizes a flash heating of the oil solvent to produce a visible vapor or smoke. The ambient fogger atomizes a tiny jet of liquid in a venturi tube through which passes an ultrahigh-velocity air stream. The materials used in fogging machines depend on the type of fogger. Thermal devices use oil only, whereas ambient generators use water, emulsions, or oils.

DUSTS

Historically dusts (D) have been the simplest formulations of pesticides and the easiest to apply. Examples of the undiluted toxic agent are sulfur dust, used on ornamentals, and one of the older household roach dusts, sodium fluoride. An example of the toxic agent with active diluent would be one of the garden insecticides having sulfur dust as its carrier or diluent. A toxic agent with an inert diluent is the most common type of dust formulation in use today, both in the home garden and in agriculture. Insecticide-fungicide combinations are applied in this manner, the carrier being an inert clay, such as pyrophyllite. The last type, the aerosol dust, is a finely ground silica in a liquefied gas propellant that can be directed into crevices of homes and commercial structures for insect control.

Despite their ease in handling, formulation, and application, dusts are the least effective and, ultimately, the least economical of the pesticide formulations. The reason is that dusts have a very poor rate of deposit on foliage, unless it is wet from dew or rain. In agriculture, for instance, an aerial application of a standard dust formulation of pesticide will result in 10 percent to 40 percent of the material reaching the crop. The remainder drifts upward and down wind. Psychologically, dusts are annoying to the nongrower who sees great clouds of dust resulting from an aerial application, in contrast to the grower who believes he or she is receiving a thorough application for the very same reason. The same statement may be relevant to the avid user of dusts in his garden and the abstaining neighbor! Under similar circumstances, an aerial or garden hand sprayer application of a water emulsion spray will deposit 50 to 80 percent of the pesticide on target.

AEROSOLS

Most of us have been raised in the aerosol culture: bug bombs, hair sprays, underarm deodorants, home deodorizers, oven sprays, window sprays, repellents, paints, garbage can and shower-tub disinfectants, and supremely, the foot and crotch sprays or antiitch remedies. In pesticides, the insecticides are dominant. Developed during World War II for the GIs, the pushbutton variety is used as space sprays to knock down flying insects. More recently the total-release aerosol has been designed to discharge its entire contents in a single application. They are available for commercial pest control operators as well as for homeowners. In either case, the nozzle is depressed and locked into place, permitting the aerosol total emission while the occupants leave and remain away for a few hours. Aerosols are effective only against resident flying and crawling insects and provide no residual effect, as do conscientiously applied sprays. How do aerosols work? Essentially, the active ingredients must be soluble in the highly volatile, liquified gas in its pressurized condition. When the liquified gas (fluorohydrocarbons) is atomized, it evaporates almost instantly, leaving the microsized droplets of toxicant suspended in air. I should point out that aerosols commonly produce droplets well below 10 microns in diameter, which are respirable. This means that they will be absorbed by alveolar tissue in the lungs rather than impinging in the bronchioles, as do larger droplets. Consequently, aerosols of all varieties should be handled with discretion and breathed as little as possible.

GRANULAR PESTICIDES

Granular (G) pesticides overcome the disadvantages of dusts in their handling characteristics. The granules are small pellets formed from various inert clays and sprayed with a solution of the toxicant to give the desired content. After the solvent has evaporated, the granules are packaged for use. Granular materials range in size from 20 to 80 mesh, a unit that refers to the number of grids per inch of screen through which they will pass. Only insecticides and a few herbicides are formulated as granules. They range from 2 to 25 percent active ingredient and are used almost exclusively in agriculture, although systemic insecticides as granules can be purchased for lawn and ornamentals. Granular materials may be applied at virtually any time of day, since they can be applied aerially in winds up to 20 mph without problems of drift, an impossible task with sprays or dusts. They also lend themselves to soil application in the drill at planting time to protect the roots from insects or to introduce a systemic to the roots for transport to above-ground parts in lawns and ornamentals.

FUMIGANTS

Fumigants are a rather loosely defined group of formulations. The plastic insecticide-impregnated strips and pet collars of the same materials are really a slow-release formulation that permits the insecticide to work its way slowly to the surface and volatilize. Moth crystals and moth balls—paradichlorobenzene and naphthalene, respectively—are crystalline solids that evaporate slowly at room temperatures, exerting both a repellent as well as an insecticidal effect. (They can also be used in small quantity to keep cats and dogs off of or away from their favorite parking places.) Soil fumigants are used in horticultural nurseries, greenhouses, and on high-value cropland to control nematodes, insect larvae, and adults and sometimes to control diseases. Depending on the fumigant, the treated solids may require covering with plastic sheets for several days to retain the volatile chemical, allowing it to exert its maximum effect.

IMPREGNATING MATERIALS

Impregnating materials mentioned here will include only treatment of woolens for mothproofing and timbers against wood-destroying organisms. For several years, woolens and occasionally leather garments have been mothproofed in the final stage of dry cleaning (using chlorinated solvents). The last solvent rinse contains an ultralow concentration of the chlorinated cyclodiene insecticide, dieldrin, which has long residual qualities against moths and leather-eating beetle larvae. Railroad ties, telephone and light poles, fence posts, and other wooden objects that have close contact with or are actually buried in the ground soon begin to deteriorate as a result of attacks from fungal decay microorganisms and insects, particularly termites, unless treated with fungicides and insecticides. Such treatment permits poles to stand for 40 to 60 years that would otherwise have been replaced in 5 to 10 years.

Impregnated Shelf Papers

Impregnated shelf papers, strips, and cords containing insecticides are in a rapid state of market decline. Thoroughly effective against stored products insect pests, they usually contained one of the chlorinated insecticides to give long residual activity. These insecticides, along with most others, cannot be used where food and food utensils are stored, according to regulations established by the Environmental Protection Agency, which have resulted in the declining use of impregnated materials.

Impregnated Wax Bars

Impregnated wax bars contain herbicides that are selective against broad-leaf plants. When dragged over grass lawns in a uniform pattern, enough is rubbed off on weeds to eliminate them, leaving the grass unaffected. This type of application is very selective, represents a spot application that is not disruptive to the environment, and is the type that should be strongly encouraged.

FERTILIZER COMBINATIONS

Fertilizer combinations are formulations fairly familiar to the urbanite who has purchased a lawn or turf fertilizer that contained a herbicide for crabgrass control, insecticides for grubs and sod webworms, or a fungicide for numerous lawn diseases. Fertilizer-insecticide mixtures have been made available to growers, particularly in the corn belt, by special order with the fertilizer distributor. The fertilizer and insecticide can then be applied to the soil during planting in a single, economical operation.

BAITS

Baits can be purchased or formulated at home. Those that are purchased contain low levels of the toxicant incorporated into materials that are relished by the target pests. Here is another example where spot application—placing the bait in selected places accessible only to the target species—permits the use of very small quantities of oftentimes highly toxic materials in a totally safe manner, with no environmental disruption.

SLOW-RELEASE INSECTICIDES

Slow-release insecticides are new and only a few are available to the homeowner. The principle, as mentioned in the paragraph on fumigants, involves the incorporation of the insecticide in a permeable covering that permits its escape at a reduced, but effective, rate. One agricultural insecticide has been encapsulated into extremely small plastic spheres, which are then sprayed on crops. The insecticide escapes through the sphere wall over an extended period, thus preserving its effectiveness much longer than if formulated as an emulsible concentrate.

The paint-on lacquer is probably the most recent innovation in structural pest control. Here the insecticide is dissolved in a special solvent containing a small quantity of dissolved plastic or lacquer. Following its paint-on application as a spot treatment in homes, restaurants, and food-handling establishments, the solvent evaporates, leaving the insecticide incorporated in the thin film of transparent lacquer. Over time the insecticide "blooms" at a constant rate presenting a freshly treated surface to crawling insects at all times. The new insecticidal adhesive tapes work essentially by the same principle but are perhaps related more to the plastic strips mentioned in the section on fumigants. The adhesive back is exposed by removing a protective strip, and the tape is attached beneath counters, under shelves, and in other protected places. These two new formulations are now available to homeowners.

CONCLUSION

In closing this chapter on formulations, it might appear that there is no limit to the different forms in which a pesticide can be prepared. This is almost the case. While reading the last entry on slow-release formulations, your imagination may have been stimulated to think about formulations of the future. If not through economy, then by edict of the EPA, we will learn to formulate and apply pesticides in extremely conservative ways, to preserve our health, our resources, and the environment, which at times appear to be in some jeopardy. In summary, pesticides are formulated to improve their properties of storage, handling, application, effectiveness, and safety.

PART II

CHEMICALS USED IN
THE CONTROL OF INVERTEBRATES:
ANIMALS WITHOUT BACKBONES

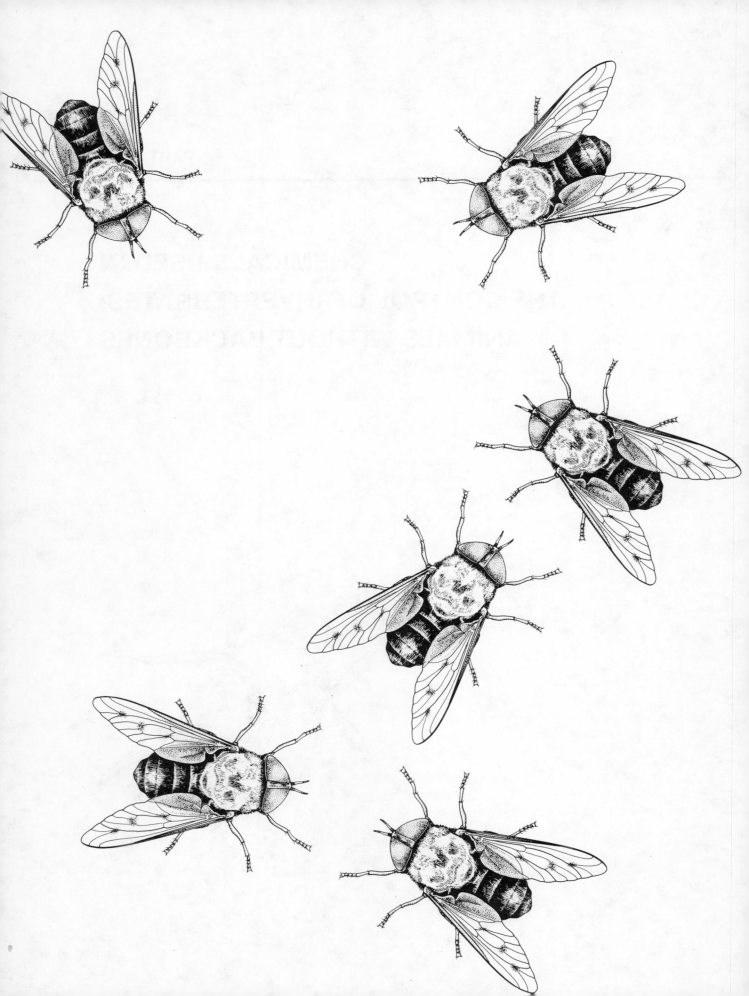

Insecticides

Humans have been on earth somewhere between 1 and 2 million years, a figure that most of us have difficulty in imagining. In contrast, however, try to visualize 250,000,000 years, the period for which insects are known to have existed. Despite their head start, humanity has been able to carve its niche and forge a path through the insect devastations. We have learned to live and compete with the insect world. There is no way to determine when insecticides began to become a tool, but we can guess that the first materials used by our primitive ancestors that were classed as *insecticides* in the crudest definition of the word were mud and dust spread over their skin to repel biting and tickling insects, resembling the habits of water buffalo, pigs, and elephants.

History doesn't tell us very much about chemicals used against insects. The earliest records of insecticides pertain to the burning of "brimstone" (sulfur) as a fumigant. Pliny the Elder (A.D. 23–79) recorded most of the earlier insecticide uses in his *Natural History*, collected largely from the folklore and Greek writings of the previous two or three centuries. Included among these were the use of gall from a green lizard to protect apples from worms and rot. In the interim, a variety of materials have been used with doubtful results: extracts of pepper and tobacco, hot water, soapy water, whitewash, vinegar, turpentine, fish oil, brine, lye, and many others.

Even as recently as 1940, our insecticide supply was still limited to several arsenicals, petroleum oils, nicotine, pyrethrum, rotenone, sulfur, hydrogen cyanide gas, and cryolite. World War II opened the Chemical Era with the introduction of a totally new concept of insecticide control chemicals—synthetic organic insecticides, the first of which was DDT.

ORGANOCHLORINES

Let's begin our story with a group of insecticides that should be familiar because of the notoriety given them by the press and environmentalists. The organochlorines are insecticides that contain carbon (thus the name *organo*-), chlorine, and hydrogen. They are also referred to by other names: *chlorinated hydrocarbons, chlorinated organics, chlorinated insecticides, chlorinated synthetics,* and perhaps others.

DDT and Related Insecticides

You are undoubtedly familiar with the action of EPA in banning all uses of DDT, effective January 1, 1973. In retrospect, DDT can now be considered the pesticide of greatest historical significance, as it affected human health, agriculture, and the environment. The story of its rise to stardom, carrying with it the Nobel Prize, and of decline to infamy is rather sensational and should be briefly narrated for the uninitiated.

Easy to say, easy to remember, DDT is probably the best known and most notorious chemical of this century. It is the most facinating, and remains to be acknowledged as the most useful insecticide developed. Surprisingly, DDT is more than 100 years old. It was first synthesized by a German graduate student in 1873, who had no idea of its tremendous insecticidal value, and after synthesis it was thrown out and forgotten. In 1939 a Swiss entomologist, Dr. Paul Müller, rediscovered DDT while searching for a long-lasting insecticide against the clothes moth. DDT proved to be extremely effective against flies and mosquitoes, ultimately bringing to Dr. Müller the Nobel Prize for Medicine in 1948 for his lifesaving discovery. We should keep in mind that its most beneficial use was in public health, for malaria control, and in Third World nations it still is so used.

More than 4 billion pounds of DDT have been used throughout the world for insect control since 1940, and 80 percent of that amount was used in agriculture. Production reached its maximum in the United States in 1961, when 160 million pounds were manufactured. The greatest agricultural benefits from DDT have been in the control of the Colorado potato beetle and several other potato insects, the codling moth on apples, corn earworm, cotton bollworm, cotton budworm, pink bollworm on cotton, and the worm complex on vegetables. It has been most useful against the gypsy moth and the spruce budworm in forests. From the standpoint of human medicine, DDT has been most successful against mosquitoes that transmit malaria and yellow fever, against body lice that can carry typhus, and against fleas that are vectors of plague.

One of the most amazing features of DDT was its low cost. Most of that sold to the World Health Organization went for less than 22 cents per pound! Without question, it was the most economical in-

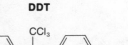

DDT

1,1,1-trichloro-2,2-bis(*p*-chlorophenyl)ethane

secticide ever sold. A federal ban on the use of DDT was declared by the Environmental Protection Agency on January 1, 1973. DDT was declared to be an environmental hazard due to its long residual life and to its accumulation, along with the metabolite DDE, in food chains, where it proved to be detrimental to certain forms of wildlife. It is no longer available to the grower or homeowner.

DDT belongs to the chemical class of diphenyl aliphatics, which means that it consists of an aliphatic, or straight carbon chain, with two (di-) phenyl rings attached, as in the illustration. DDT was first known chemically as dichloro diphenyl trichloroethane, hence DDT.

The chemical structure for DDT is presented here for the reader, more as a matter of historical curiosity than any other. It will suffice to say that there are five relatives of DDT that should be mentioned, because they have all had an early role in pest control: TDE (or DDD), methoxychlor, Perthane, dicofol, and chlorobenzilate. The latter two are not really insecticides, but rather are acaricides (miticides). More information is given on these and other pesticides in Appendix A.

Now, regarding the structure of DDT. Structures really do not reveal anything about chemical stability or persistence, but only DDT and TDE of the six listed have this good (or bad) characteristic, depending on your viewpoint. The term *persistence*, as used here, implies a chemical stability giving the products long lives in soil and aquatic environments and in animal and plant tissues. They are not readily broken down by microorganisms, enzymes, heat, or ultraviolet light. From the insecticidal viewpoint, these are good characteristics. From the environmental viewpoint, they are not. Using these qualities, the remaining DDT relatives would be considered nonpersistent.

How does DDT kill? The mode of action, or type of biological activity, has never been clearly worked out for DDT or any of its relatives. It does affect the neurons or nerve fibers in a way that prevents normal transmission of nerve impulses, both in insects and mammals. Eventually the neurons fire impulses spontaneously, causing the muscles to twitch; this may lead to convulsions and death. There are several valid theories for DDT's mode of action, but none has been clearly proved. It is sufficient to say that DDT in some complex manner destroys the delicate balance of sodium and potassium within the neuron, thereby preventing it from conducting impulses normally.

Right at the outset, I should relate a few salient points concerning DDT that aid in understanding some of the well-documented evils attributed to it. The first point is DDT's chemical stability. As mentioned before, it is very persistent, relatively stable to the ultraviolet of sunlight, not readily broken down by microorganisms in the soil or elsewhere, stable to heat and acids, and unyielding in the presence of almost all enzymes. In other words, it is poorly biodegradable. The second point is its solubility in water—zero. DDT has been reported in the chemical literature to be probably the most water-insoluble compound ever synthesized. Its water solubility is actually somewhere in the neighborhood of six parts per billion parts of water (ppb). On the other hand, it is quite soluble in fatty tissue, and, as a consequence of its resistance to metabolism, it is readily stored in fatty tissue of any animal ingesting DDT alone or DDT dissolved in the food it eats, even when it is part of another animal.

If it is not readily metabolized and thus not excreted, and if it is freely stored in body fat, it should come as no surprise that it accumulates in every animal that preys on other animals. It also accumulates in animals that eat plant tissue bearing even traces of DDT. Here we aim at the dairy and beef cow. The dairy cow excretes (or secretes) a large share of the ingested DDT in its fat. Humans drink milk and eat the fatted calf. Guess where the DDT is now.

The same story is repeated time and again in food chains ending in the osprey, falcon, golden eagle, seagull, pelican, and so on.

The principle of these food chain oddities is this: Any chemical that possesses the characteristics of stability and fat solubility will follow the same biological magnification (condensed to *biomagnification*) as DDT. One group of chemicals, which has no insecticidal properties, the polychlorinated biphenyls, have over the years been slowly released into the environment, and as a consequence of these two magic characteristics, have climbed the food chain just as DDT has. Other insecticides incriminated to some extent in biomagnification, belonging to the organochlorine group, are TDE, DDE (a major metabolite of DDT), dieldrin, aldrin, several isomers of BHC, endrin, heptachlor, and mirex. And, of course, they all possess these two crucial prerequisites.

The rise and fall of DDT contains a lesson, and this is perhaps as good a place as any to moralize. Because of DDT's great success in World War II against body lice in Naples, during the typhus outbreak, and in the Pacific, against mosquitoes known to vector malaria, after the war it was rapidly swept into agricultural use with inadequate basic knowledge. And, because of its effectiveness against a host of agricultural insect pests and its ridiculously low cost, it was overused, and abused, and then—when it caused problems—was banned in rather a panic. The emotional issues intertwined in the newspaper and street trials that ensued narrowly missed the magnitude of importance shared only by the flag, motherhood, and perhaps apple pie.

The lesson we can learn from this is: There is an absolute need for an informed, cautious, and—according to available knowledge—correct way of employing a specific chemical for pest control (of course, it does not have to be an insecticide); it is quite likely that new "DDT cases" are in the making; and, consequently, there is the necessity for basic research performed in autonomous institutions not subjected to the need for competing in the economic marketplace. And, because this research is basic, as opposed to applied, it will of necessity be slow, expensive, long-term and not immediately applicable.

So, let that be a lesson!

BHC or Benzenehexachloride

HCH (BHC), hexachlorocyclohexane, was first discovered in 1825. But, like DDT, it was not known to have insecticidal properties until 1940, when French and British entomologists found the material to be active against all insects tested.

It is made by chlorinating benzene, which results in a product made up of several isomers, that is, molecules containing the same kinds and numbers of atoms but differing in the internal arrangement of those atoms. HCH, for instance, has five isomers, named, after the Greek letters, *Alpha, Beta, Gamma, Delta,* and *Epsilon.* After much laboratory work in isolating and identifying these isomers, the chemists found to their great surprise that only the Gamma isomer had insecticidal properties. In a normal mixture of HCH, the Gamma isomer comprises only about 12 percent of the total, leaving the other four isomers as inert material or insecticidally inactive ingredients.

HCH (BHC)

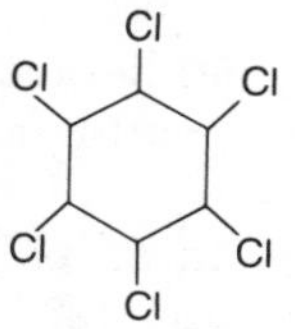

1,2,3,4,5,6-hexachlorocyclohexane

Since the Gamma isomer was the only active ingredient, methods were developed to manufacture a product, lindane, containing 99 percent Gamma isomer, which was effective against most insects, but also quite expensive, making it impractical for crop use.

Technical grade HCH has one highly undesirable characteristic, a prominent musty odor and flavor, which helps in remembering several other points of interest. This odor is from the inert isomers, which are more persistent than the odorless Gamma isomer in animal and plant tissues as well as in soil. As a result, root and tuber crops planted in soils previously treated with HCH retained its odor and were usually unsalable. There were reports of the same problems with leafy vegetables, poultry, eggs, and milk that directly or indirectly, came in contact with HCH residues.

The mode of action for lindane is also not completely understood, but its effects on insects and mammals superficially resemble those of DDT and are probably brought on by a sodium-potassium imbalance in the neurons. It is known to be a neurotoxicant whose effects are normally seen within hours and result in increased activity, tremors, and convulsions leading to prostration.

Lindane was odorless and had a high degree of volatility. It quickly became popular as a household fumigant sold as pellets to be attached to light bulbs or to small, decorative electric wall vaporizers. These were later found to be hazardous to humans and their pets and were removed from the market.

Cyclodienes

The cyclodienes are a prominent and extremely useful group of insecticides, also known as the *diene-organochlorine insecticides*.

They were developed after World War II and are therefore of more recent origin than DDT (1939) and HCH (1940). The eight compounds listed as follows were first described in the scientific literature or patented in the year indicated: chlordane, 1945; aldrin and dieldrin, 1948; heptachlor, 1949; endrin, 1951; mirex, 1954; endosulfan, 1956; and Kepone, 1958. Other cyclodienes developed in the United States and Germany are of minor importance in a general survey. These include Isodrin, Alodan, Bromodan, and Telodrin.

Generally, the cyclodienes are persistent insecticides and are stable in soil and relatively stable to the ultraviolet action of sunlight. Consequently, they have been used in greatest quantity as soil insecticides (especially chlordane, heptachlor, aldrin, and dieldrin), for the control of termites and soil-borne insects whose immature stages (larvae) feed on the roots of plants. Because of their persistence, the use of cyclodienes on crops was restricted; undesirable residues remained beyond the time for harvest. To suggest the effectiveness of cyclodienes as termite control agents, structures treated with chlordane, aldrin, and dieldrin in the year of their development are still protected from damage. This is 33 and 30 years, respectively. It would be elementary to say that these insecticides are the most effective, long-lasting, economical, and safest termite control agents known. However, several other soil insects have become resistant to these materials in agriculture, and this has resulted in a decline of their use.

The most valuable, and produced in the greatest quantity, were chlordane and dieldrin. Their structures are presented as typical of the cyclodienes.

CHLORDANE

1,2,4,5,6,7,8,8-octachloro-3a,4,7,7a-tetrahydro-
4,7-methanoindane

DIELDRIN

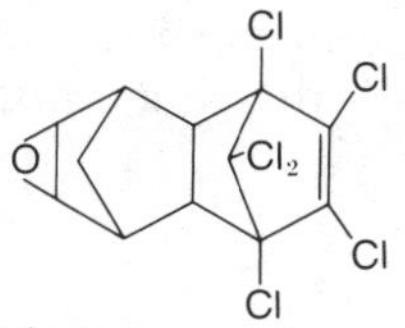

endo-exo

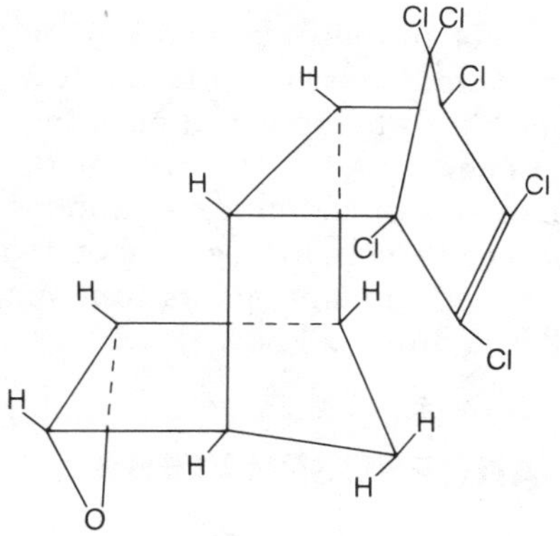

1,2,3,4,10,10-hexachloro-6,7-epoxy-
1,4,4a,5,6,7,8,8a-octahydro-1,4-*endo-exo*-
5,8-dimethanonaphthalene

The nomenclature and chemistry of the cyclodienes is rather complicated and will not be studied in depth. It is of value to mention that the cyclodienes do have three-dimensional structures and thus possess stereoisomers, again having the same kinds and numbers of atoms but having atoms that differ in their spatial location and structure.

The cyclodienes have about equal toxicity or toxic effects on insects, mammals, and birds. They are, however, much more toxic to fish and for a very good reason. Fish are totally surrounded when the compound is introduced into water. Figuratively speaking, they eat, sleep, and breathe their aquatic environment and any toxic compound contained therein.

The modes of action of the cyclodienes are not clearly understood. It is known that they are neurotoxicants that have effects similar to those of DDT and HCH. They appear to affect all animals in generally the same way, first with nervous activity followed by tremors, convulsions, and prostration. The cyclodienes undoubtedly molest the delicate balance of sodium and potassium within the neuron but in a way differing from that of DDT and HCH.

Polychloroterpene Insecticides

There are only two polychloroterpene materials, toxaphene, discovered in 1947, and strobane, introduced in 1951. Neither have ever been considered urban insecticides. Toxaphene is manufactured by the chlorination of camphene, a pine tree derivative. Toxaphene has by far the greatest use, which, surprisingly, is almost exclusively on cotton. Alone, it has a low order of toxicity to insects and is thus formulated with other insecticides. These materials are semipersistent in the soil and disappear in 3 to 4 weeks from the surfaces of most plant tissue. They are fairly easily metabolized by mammals

TOXAPHENE

chlorinated camphene containing 67 to 69 percent chlorine

PHOSPHORIC ACID

and are not stored in body fat to any great extent, as are DDT, HCH, or the cyclodienes. Despite the low insect toxicity, fish are highly susceptible to toxaphene poisoning, in the same magnitude as the cyclodienes.

The modes of action for toxaphene and strobane are similar to the cyclodiene insecticides, acting on the neurons and causing an imbalance in sodium and potassium ions. Beyond this generalization, really very little is known.

ORGANOPHOSPHATES

This next group, the chemically unstable organophosphate (OP) insecticides, has virtually replaced the persistent organochlorine compounds. This is especially true with regard to their use around the home and garden.

The OPs have several commonly used names, any of which are correct. Organic phosphates, phosphorus insecticides, nerve gas relatives, phosphates, phosphate insecticides, and phosphorus esters or phosphoric acid esters. They are all derived from phosphoric acid and are generally the most toxic of all pesticides to vertebrate animals. Because of their chemical structure and mode of action, they are related to the "nerve gases." Their insecticidal action was observed in Germany during World War II in the study of materials closely related to the nerve gases sarin, soman, and tabun. Initially, the discovery was made in search of substitutes for nicotine, which was in critically short supply in Germany.

The OPs have two distinctive features. First, they are generally much more toxic to vertebrates than are the organochlorine insecticides, and, second, they are chemically unstable or nonpersistent. It is this latter quality that brings them onto the agricultural scene as substitutes for the persistent organochlorines, particularly DDT.

We can discuss the mode of action of this gigantic group with considerable confidence. The OPs exert their toxic action by tying up or inhibiting certain important enzymes of the nervous system, cholinesterases (ChE). Throughout the nervous system in vertebrates as well as insects, are electrical switching centers, or synapses, where the electrical signal is carried across a gap to a muscle or another nerve fiber (neuron) by a chemical, in many instances acetylcholine (ACh). After the electrical signal (nerve impulse) has been conducted across the gap by ACh, the ChE enzyme moves in quickly and removes the ACh so the circuit will not be "jammed." These chemical reactions happen extremely rapidly and go on constantly under normal conditions. When OPs enter the scene, they attach to the ChE in a way that prevents them from removing the ACh, and the circuits jam because of the accumulation of ACh. What this really says is that the accumulation of ACh interferes with the neuromuscular junction, producing rapid twitching of voluntary muscles and finally paralysis. This process is of particular importance in proper functioning of the respiratory system.

Normally *organophosphate* is used as a generic term to include all of the insecticides containing phosphorus. Now we will consider the atoms attached to the phosphorus. OPs that are combinations of different alcohols and different phosphorus acids are termed *esters*.

Esters of phosphorus have varying combinations of oxygen, carbon, sulfur, and nitrogen attached to the phosphorus and so have different identities. Below are shown the six subclasses of OPs, only to help explain some of the seemingly odd chemical names given to these insecticides.

As with the last two groups of insecticides, the OPs are further divided into three classes, the aliphatic, phenyl, and heterocyclic derivatives. Each class has several materials to be examined.

PHOSPHATE   **PHOSPHONATE**

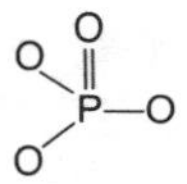

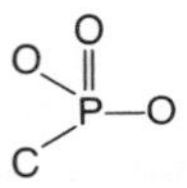

PHOSPHOROTHIOATE 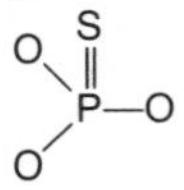   **PHOSPHOROTHIOLATE**

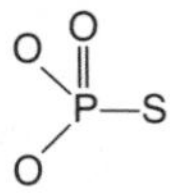

PHOSPHORODITHIOATE 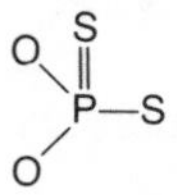   **PHOSPHORAMIDATE**

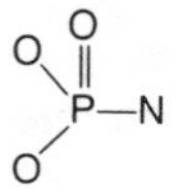

Aliphatic Derivatives

The term aliphatic literally means "carbon chain," and the linear arrangement of carbon atoms differentiates them from ring or cyclic structures. All of the aliphatic OPs are simple phosphoric acid derivatives bearing short carbon chains.

The first OP introduced into agriculture was TEPP in 1946. It is the only useful pyrophosphate and is probably the most toxic. It was never available for home use and is presented here out of academic interest. Because TEPP is very unstable in water, it hydrolyzes (breaks down) quickly after spraying on crops and disappears within 10 to 12 hours.

PYROPHOSPHATE

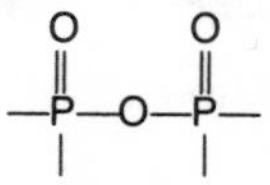

TEPP

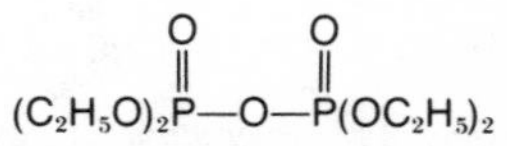

$(C_2H_5O)_2P—O—P(OC_2H_5)_2$

tetraethyl pyrophosphate

MALATHION

$$\begin{array}{c} S \\ \parallel \end{array} \quad \begin{array}{c} O \\ \parallel \\ CH_2-C-OC_2H_5 \end{array}$$

$$(CH_3O)_2P-S-CH-C-OC_2H_5$$

$$\begin{array}{c} \parallel \\ O \end{array}$$

diethyl mercaptosuccinate, *S*-ester with
O,O-dimethyl phosphorodithioate

TRICHLORFON (Dylox®)

$$\begin{array}{cc} O & H \\ \parallel & \mid \\ (CH_3O)_2P-CHCCl_3 \end{array}$$

dimethyl (2,2,2-trichloro-1-hydroxyethyl)
phosphonate

MONOCROTOPHOS (Azodrin®)

$$\begin{array}{ccc} O & CH_3 & O \\ \parallel & \mid & \parallel \\ (CH_3O)_2P-O-C=CHC-NH-CH_3 \end{array}$$

3-hydroxy-*N*-methyl-*cis*-crotonamide
dimethyl phosphate

DICHLORVOS (Vapona®)

$$\begin{array}{c} O \\ \parallel \\ (CH_3O)_2P-O-CH=CCl_2 \end{array}$$

2,2-dichlorovinyl dimethyl phosphate

MEVINPHOS (Phosdrin®)

$$\begin{array}{ccc} O & CH_3 & O \\ \parallel & \mid & \parallel \\ (CH_3O)_2P-O-C=CHC-OCH_3 \end{array}$$

methyl 3-hydroxy-*alpha*-crotonate,
dimethyl phosphate

Malathion, however, is one of the safest of the organophosphates and is commonly used in and around the home with little or no hazard either to humans or their pets.

Trichlorfon is a chlorinated OP, which has been useful for crop pest control and fly control around barns and other farm buildings.

Monocrotophos is a relatively new aliphatic OP containing nitrogen. It is a plant-systemic insecticide, but it has had limited use in agriculture because of its high mammalian toxicity and is not available to the homeowner.

Systemic insecticides are those that are taken into the roots of plants and translocated to the above-ground parts, where they are toxic to any sucking insects feeding on the plant juices. Normally caterpillars and other plant tissue-feeding insects are not controlled, because they do not ingest enough of the systemic containing juices to be affected.

Contained among the aliphatic derivatives are several plant systemics, dimethoate, dicrotophos, oxydemetonmethyl, and disulfoton, all of which can be used safely by the homeowner.

DIMETHOATE (Cygon®)

$$\begin{array}{cc} S & O \\ \parallel & \parallel \\ (CH_3O)_2P-S-CH_2C-NH-CH_3 \end{array}$$

O,O-dimethyl *S*-(*N*-methylcarbamoylmethyl)
phosphorodithioate

OXYDEMETONMETHYL (Meta Systox®)

$$\begin{array}{cc} O & O \\ \parallel & \parallel \\ (CH_3O)_2P-S-CH_2CH_2-S-C_2H_5 \end{array}$$

S-[2-(ethylsulfinyl)ethyl] *O,O*-dimethyl
phosphorothioate

DICROTOPHOS (Bidrin®)

$$\begin{array}{ccc} O & CH_3 & O \\ \parallel & \mid & \parallel \\ (CH_3O)_2P-O-C=CHC-N(CH_3)_2 \end{array}$$

3-hydroxy-*N,N*-dimethyl-*cis*-crotonamide
dimethyl phosphate

DISULFOTON (Di-Syston®)

$$\begin{array}{c} S \\ \parallel \\ (C_2H_5)_2P-S-CH_2CH_2-S-C_2H_5 \end{array}$$

O,O-diethyl *S*-2-[(ethylthio)ethyl]
phosphorodithioate

Dichlorvos is an aliphatic OP with a very high vapor pressure, giving it strong fumigant qualities. It has been incorporated into vinyl plastic pet collars and pest strips, from which it is released slowly. It lasts several months and is useful for insect control in the home and other closed areas.

Mevinphos is a highly toxic OP used in commercial vegetable production because of its very short insecticidal life. It can be applied up to one day before harvest for insect control, yet it leaves no residues on the crop to be eaten by the consumer.

In summary, the aliphatic organophosphate insecticides are the simplest in structure of the organophosphate molecules. They have a wide range of toxicities, and several possess a relatively high water solubility, giving them plant-systemic qualities several of which are useful around the home.

Phenyl Derivatives

You will recall from the section on chemistry that the benzene ring, when attached to other groups, is referred to as *phenyl*. The phenyl OPs contain a benzene ring with one of the ring hydrogens displaced by attachment to the phosphorus moiety and others frequently displaced by Cl, NO_2, CH_3, CN, S, and so on. The phenyl OPs are generally more stable than the aliphatic OPs; consequently their residues are longer lasting.

Parathion is the most familiar of the phenyl OPs, being, in 1947, the second phosphate insecticide introduced into agriculture. The first, TEPP, was introduced in 1946. As a result of its age and utility, parathion's total usage is greater than that of many of the less useful materials combined. Ethyl parathion was the first phenyl derivative used commercially and, because of its hazard, has not been available to the homeowner.

Methyl parathion became available in 1949 and proved to be more useful than (ethyl) parathion because of its lower toxicity to humans and domestic animals and broader range of insect control. Its shorter residual life also makes it more desirable in certain instances. This material is also not used by the layperson.

Systemic insecticides are also found in the phenyl OPs. They are, however, usually animal systemics used for the control of the cattle grub; ronnel and crufomate are examples.

ETHYL PARATHION

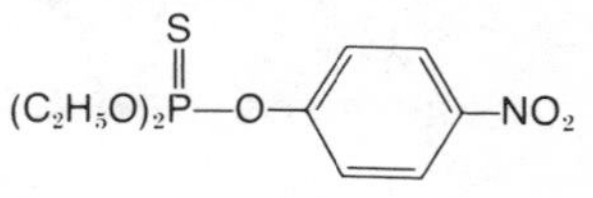

O,O-diethyl O-p-nitrophenyl phosphorothioate

METHYL PARATHION

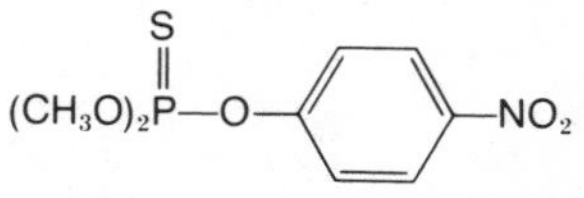

O,O-dimethyl O-p-nitrophenyl phosphorothioate

RONNEL (Korlan®)

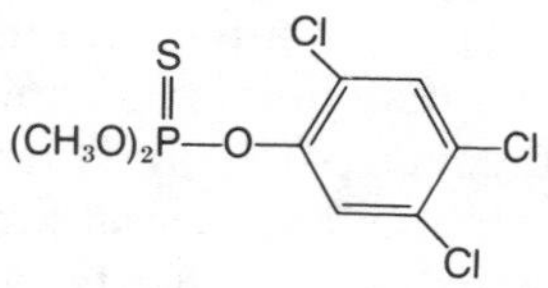

O,O-dimethyl O-2,4,5-trichlorophenyl phosphorothioate

CRUFOMATE (Ruelene®)

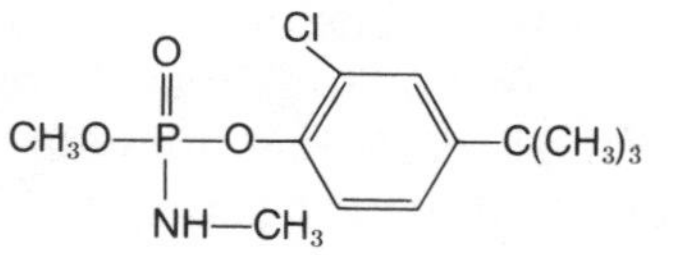

4-tert-butyl-2-chlorophenyl methyl methylphosphoramidate

GARDONA®

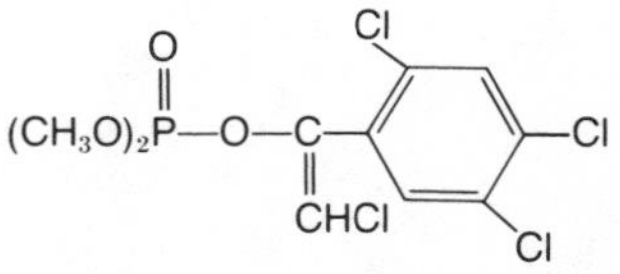

2-chloro-1-(2,4,5-trichlorophenyl) vinyldimethyl phosphate

Gardona is a home-safe OP much like malathion in its overall usefulness against home and garden pests.

DIAZINON

$(C_2H_5)_2$P—O

O,O-diethyl *O*-(2-isopropyl-4-methyl-6-pyrimidyl)
phosphorothioate

AZINPHOSMETHYL (Guthion®)

$(CH_3O)_2$P—S—CH$_2$

O,O-dimethyl *S*(4-oxo-1,2,3-benzotriazin-
3(4*H*)-ylmethyl) phosphorodithioate

CHLORPYRIFOS (Dursban®)

$(C_2H_5O)_2$P—O

·diethyl *O*-(3,5,6-trichloro-2-pyridyl)
phosphorothioate

Heterocyclic Derivatives

The term *heterocyclic* means that the ring structures are composed of different or unlike atoms. In a heterocyclic compound, for example, one or more of the carbon atoms is displaced by oxygen, nitrogen or sulfur, and the ring may have three, five, or six atoms.

The first (1952) insecticide made available in this group was probably diazinon. Note that the six-membered ring contains two nitrogen atoms, very likely the source of its proprietary name, since one of the constituents used in its manufacture is pyrimidine, a diazine.

Diazinon is a relatively safe OP that has an amazingly good track record around the home. It has been effective for practically every conceivable use: insects in the home, lawn, garden, ornamentals, around pets, and for fly control in stables and pet quarters.

Azinphosmethyl is the second oldest member of this group (1954) and is used in U.S. agriculture. It serves both as an insecticide and acaricide in cotton production and is not available to the layperson.

Chlorpyrifos has become the most frequently used insecticide by pest control operators in homes and restaurants for controlling cockroaches and other household insects.

In closing, the heterocyclic organophosphates are complex molecules and generally have longer-lasting residues than many of the aliphatic or phenyl derivatives. Also, because of the complexity of their molecular structures, their breakdown products (metabolites) are frequently many, making their residues sometimes difficult to measure in the laboratory. Consequently, their use by growers on food crops is somewhat less than either of the other two groups of phosphorus-containing insecticides.

ORGANOSULFURS

The organosulfurs, as the name suggests, have sulfur as their central atom. They resemble the DDT structure in that most have two phenyl rings.

Dusting sulfur alone is a good acaricide (miticide), particularly in hot weather. The organosulfurs, however, are far superior, requiring much less material to achieve control. You may quickly surmise that sulfur in combination with phenyl rings is particularly toxic to mites, and your observation is accurate. Of greater interest, however, is that the organosulfurs have very low toxicity to insects. As a result, they are used only for mite control.

This group has one other valuable property: They are usually ovicidal as well as being toxic to the young and adult mites.

Tetradifon is one of the older acaricides and typically bears the sulfur and twin phenyl rings, as do most of the organosulfurs.

No doubt the oldest of this group is Aramite, introduced in 1951.

TETRADIFON (Tedion®)

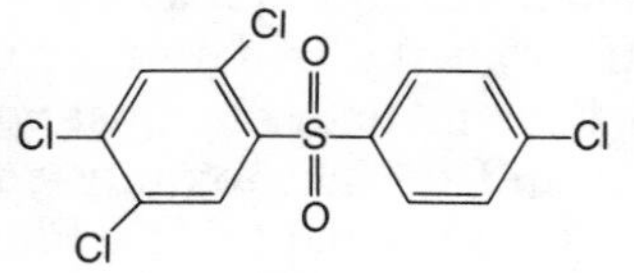

p-chlorophenyl 2,4,5-trichlorophenyl sulfone

GENITE®

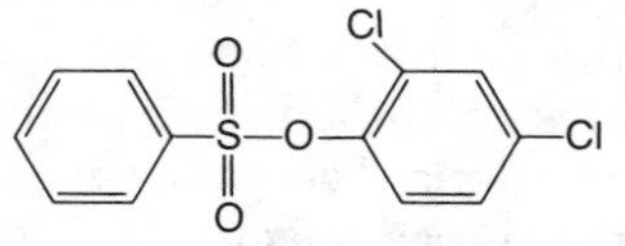

2,4-dichlorophenyl benzenesulfonate

OVEX (Ovotran®)

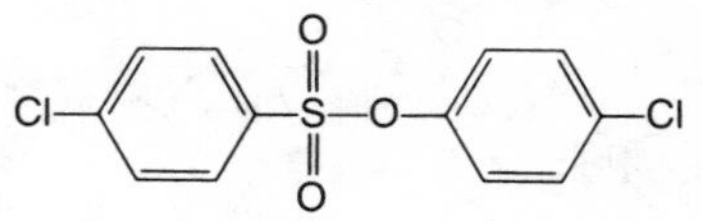

p-chlorophenyl p-chlorobenzenesulfonate

ARAMITE®

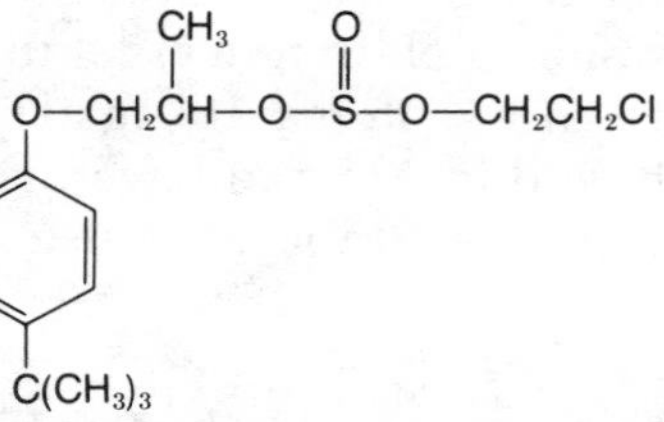

2-(p-tert-butylphenoxy)-1-methylethyl
2-chloroethyl sulfite

You may have noticed that Aramite has only one phenyl ring and is, therefore, an exception to the general rule that organosulfurs have two phenyl rings.

CARBAMATES

Since the organophosphate insecticides are derivatives of phosphoric acid, then the carbamates must be derivatives of carbamic

acid HO—C—NH$_2$. And, like the organophosphates, the mode of action of the carbamates is that of inhibiting the vital enzyme, cholinesterase (ChE).

In 1951 the carbamate insecticides were introduced by the Geigy Chemical Company in Switzerland. They fell by the wayside because the first ones were not very effective, while being quite costly. The early carbamates are presented as follows.

ISOLAN　　　**DIMETAN** 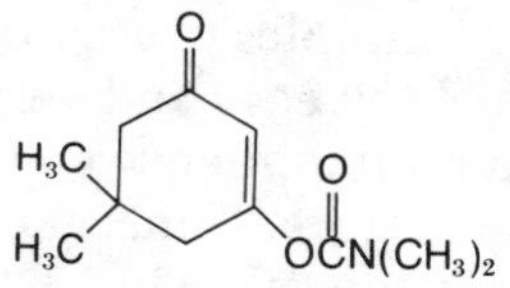　　　**PYRAMAT** 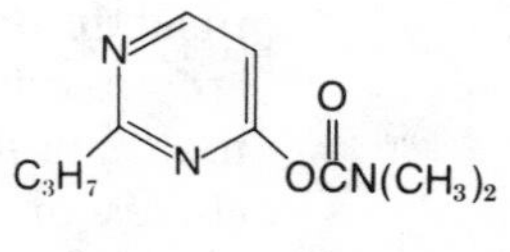　　　**PYROLAN**

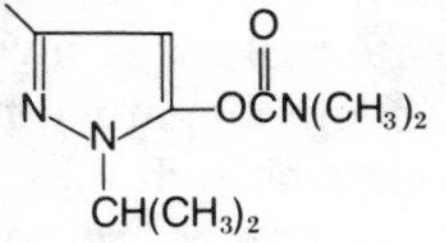

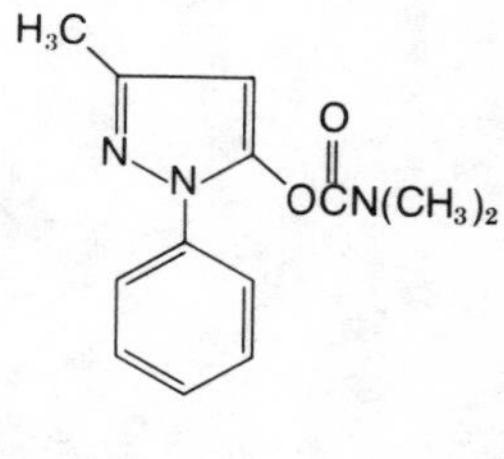

At that time it was not known that the N-dimethyl carbamates, as shown in the preceding structures, were generally less toxic to insects than the N-methyl carbamates, which were developed later and which make up the bulk of the currently used materials.

CARBARYL (Sevin®)

1-naphthyl methylcarbamate

METHOMYL (Lannate®, Nudrin®)

$$CH_3-C=N-O-C-NH-CH_3$$
$$S-CH_3$$

methyl *N*-[(methylcarbamoyl)oxy]thioacetimidate

ALDICARB (Temik®)

$$CH_3-S-CCH=N-O-C-NH-CH_3$$

2-methyl-2-(methylthio) propionaldehyde
O-(methylcarbamoyl) oxime

CARBOFURAN (Furadan®)

2,3-dihydro-2,2-dimethyl-7-benzofuranyl
methylcarbamate

PROPOXUR (Baygon®)

o-isopropoxyphenyl methylcarbamate

Carbaryl, the first successful carbamate, was introduced in 1956. More of it has been used worldwide than all the remaining carbamates combined. Two distinct qualities have made it the most popular material: very low mammalian oral and dermal toxicity and a rather broad spectrum of insect control. This has led to its wide use as a lawn and garden insecticide. Notice that carbaryl is an N-methyl carbamate.

Methomyl is a more recently developed carbamate that has been extremely useful, especially for worm control on vegetables.

Several of the carbamates are plant systemics, indicating that they have a rather high water solubility, in order to be taken into the roots or leaves. They are also not readily metabolized by the plants. Methomyl, aldicarb, and carbofuran have distinct systemic characteristics, but are not registered for home use.

Another carbamate, Baygon or propoxur, is highly effective against cockroaches that have developed resistance to the organochlorines and organophosphates. Propoxur is used by most structural pest control operators for roaches and other household insects in restaurants, kitchens, and homes. It is also formulated in bottled sprays for home use.

In summary, the carbamates are inhibitors of cholinesterase, are plant systemics in several instances, and are, for the most part, narrow-spectrum insecticides (control a limited group of insects).

FORMAMIDINES

The formadines comprise a very new, small, but promising group of insecticides. Two examples are chlordimeform and amitraz. They are effective against the eggs and very young caterpillars of several moths of agricultural importance and are also effective against most stages of mites and ticks. Thus, they are classed as ovicides, insecticides, and acaricides. Late in 1976, chlordimeform was removed from the market by its manufacturers, CIBA-GEIGY Corporation, and Nor-Am Agricultural Products, Inc., because it proved to be carcinogenic to a cancer-prone strain of laboratory mice during high-level, lifetime feeding studies. In 1978 it was returned for use on cotton, but under very strict application restrictions.

CHLORDIMEFORM (Galecron®, Fundal®)

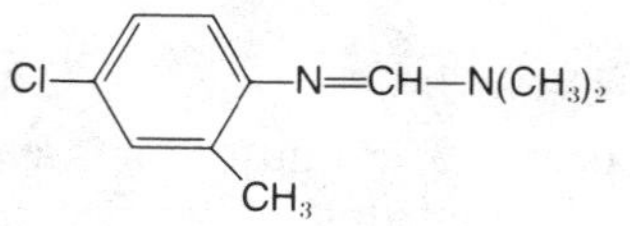

N'-(4-chloro-*o*-tolyl)-*N,N*-dimethylformamidine

AMITRAZ (Baam®, U-36059, Tactic®)

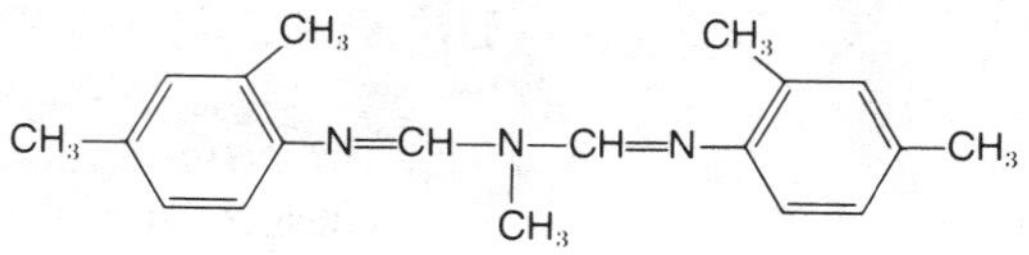

N-methyl-*N'*-2,4-xylyl-*N*-(*N*-2,4-xylylformimidoyl) formamidine

Their present value lies in the control of organophosphate- and carbamate-resistant pests. Poisoning symptoms are distinctly different from other materials. It has been currently proposed that one possible mode of action is the inhibition of a previously unmentioned enzyme, monoamine oxidase. This results in the accumulation of compounds termed *biogenic amines,* whose actions are not fully understood. However, they may act in certain instances as chemical transmitters of synapses, similar to acetylcholine. Thus the formamidines introduce a new mode of action for the insecticides and acaricides. This fact alone makes them extremely useful, for we are slowly losing ground in the battle of insect resistance to the modes of action of the older insecticide groups.

THIOCYANATES

Thiocyanates have easily recognized structural formulas. Remembering that *theion* is the Greek word for 'sulfur' and that the cyanides or cyanates end in *-CN,* we have molecules that bear *-SCN,* or *thiocyanate* endings. These insecticides have very distinct, creosotelike odors, are relatively safe to use around humans and animals, and give astonishingly quick knockdown of flying insects. Their mode of action is somewhat complex and can be said simply to interfere with cellular respiration and metabolism. These materials may be found in aerosols to be used around horses and other farm animals.

LETHANE 60®

$$CH_3(CH_2)_{10}\overset{\displaystyle O}{\overset{\|}{C}}—OCH_2CH_2—SCN$$

2-thiocyanoethyl laurate

LETHANE 384®

$$\begin{array}{l}CH_2CH_2—O—CH_2CH_2—SCN\\ |\\ O—CH_2CH_2CH_2CH_3\end{array}$$

2-(2-butoxyethoxy)ethyl thiocyanate

THANITE®

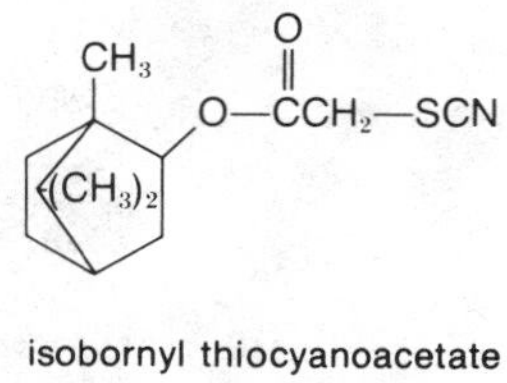

isobornyl thiocyanoacetate

DINITROPHENOLS

The dinitrophenols are mentioned only as a matter of academic interest and to make this book relatively complete. They are another group possessing easily recognized structural formulas:

PHENOL

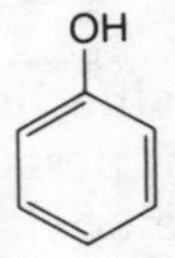

OH

or easier yet, dinitrophenol:

OH

O_2N NO_2

Di (two) nitro (NO_2) phenol

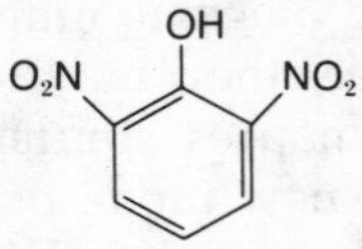

The basic dinitrophenol molecule has a broad range of toxicities. Compounds derived from it are used as herbicides, insecticides, ovicides, and fungicides. They act by uncoupling oxidative phosphorylation or basically by preventing the utilization of nutritional energy. In the 1930s, certain dinitrophenols were given by uninformed physicians to their overweight patients to induce rapid weight loss. They were extremely effective, but quite toxic, and their use resulted in several widely publicized deaths.

The oldest of this group is DNOC (3,5-dinitro-o-cresol), introduced as an insecticide in 1892. DNOC has also been used as an ovicide, herbicide, fungicide, and blossom-thinning agent. Its use has declined today to herbicidal applications where all plants are to be killed.

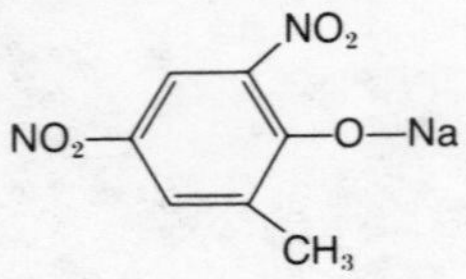

Morocide®, or binapacryl, is used exclusively as an acaricide and was introduced in 1960. Dinocap was developed in 1949 as an acaricide and fungicide and is one of the rare materials made up of several related molecular structures, only one of which is shown. Dinocap is particularly effective against powdery mildew fungi. Owing to its safety to green plants, it has often replaced the phytotoxic sulfur that is so effective against powdery mildews.

DINITROCRESOL (DNOC)

NO_2

NO_2 O—Na

CH_3

BINAPACRYL (Morocide®)

NO_2

NO_2 O—CCH=C(CH_3)$_2$

$CH_3CH_2CHCH_3$

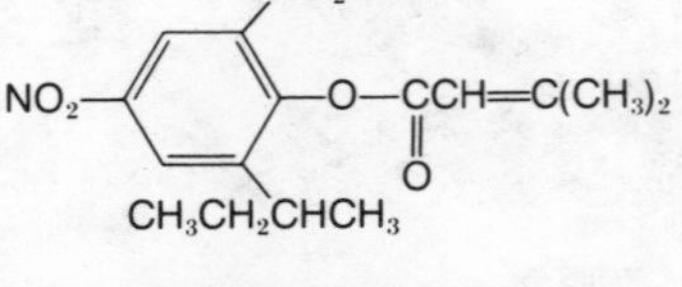

2-sec-butyl-4,6-dinitrophenyl
3-methyl-2-butenoate

DINOCAP (Karathane®)

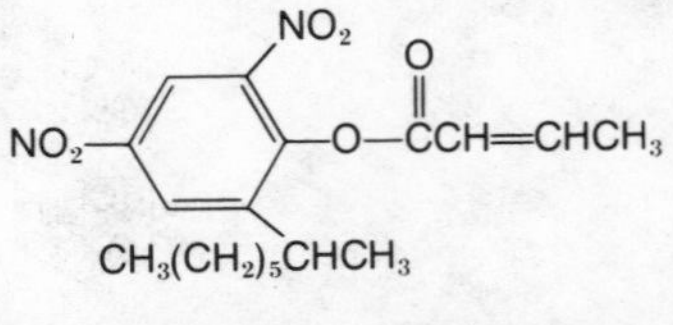

NO_2 O

NO_2 O—CCH=CHCH$_3$

$CH_3(CH_2)_5CHCH_3$

2-(1-methylheptyl)-4,6-dinitrophenyl crotonate

In summary, it can be said that the dinitrophenols have been used as pesticides in practically all classifications: ovicides, insecticides, acaricides, herbicides, fungicides, and blossom-thinning agents.

ORGANOTINS

The only reason for this short section is to introduce a relatively new group of acaricides, which double as fungicides, as you will see later. Of particular interest here is Plictran®, one of the most selective acaricides presently known, introduced in 1967. The mode of action of this group is not completely known but is believed to be the inhibition of oxidative phosphorylation at the site of dinitrophenol uncoupling. This inhibition reduces substantially, and in the case of mites, fatally, the availability of energy in the form of adenosine triphosphate (ATP). These trialkyl tins also inhibit photophosphorylation in chloroplasts and can thus serve as algicides.

tricyclohexylhydroxytin

SYNERGISTS OR ACTIVATORS

Synergists are not in themselves considered toxic or insecticidal, but are materials used with insecticides to synergize or enhance the activity of the insecticides. They are added to certain insecticides in the ratio of 8:1 or 10:1, synergist:insecticide. The first synergist was introduced in 1940 to increase the effectiveness of the plant-derived insecticide, pyrethrin. Since then many materials have been introduced, but only a few have survived, because of cost and ineffectiveness. Synergists are found in practically all of the "bug-bomb" aerosols to enhance the action of the fast knockdown insecticide, pyrethrin, against flying insects.

They were initially developed for use with pyrethrin, but have since been observed to synergize some, but not all, organophosphates, organochlorines, carbamates, as well as a few of the botanicals, or plant-derived, insecticides. Several years of mystery surrounded the mode of action of synergists, because, to further confuse matters, these compounds produced no effect with some of the insecticide classes just mentioned and even antagonized the action of others. In other words, synergists could result in a plus, neutral, or negative effect when added to insecticides. It has been well established that the synergist mode of action is the inhibition of mixed-function oxidases, enzymes that metabolize foreign compounds. In the case of insecticides, this metabolism leads either to detoxication or activation. Thus, if the inhibited enzyme normally detoxifies the insecticide, the insecticide is left intact to exert its effectiveness, and appears to be synergized. If, on the other hand, the inhibited enzyme normally activates the insecticide, as with some phosphorothioates, the insecticide is not activated, and appears to be inhibited or antagonized in its effectiveness.

The most popular synergists belong to only two molecular groups, or moieties. The first is the methylenedioxyphenyl moiety. The R_1 and R_2 are radicals, carbon chains or other groups of varying combinations, depending on the total molecular structure.

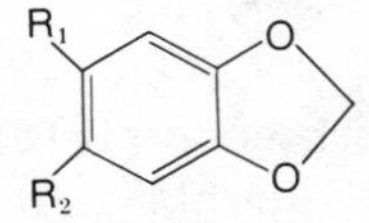

The second synergistic moiety does not have a single name but is characterized by either of the following structures.

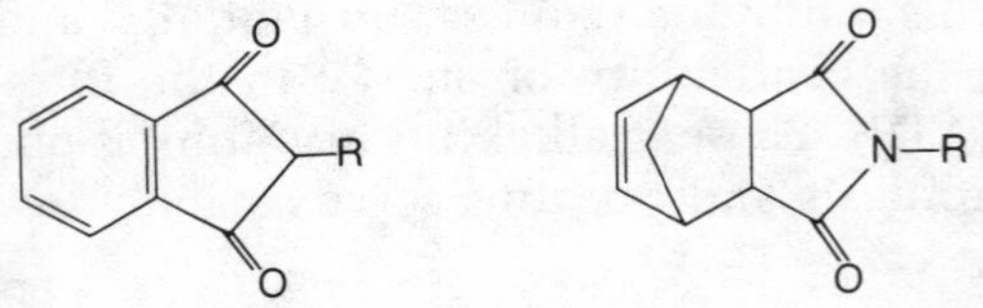

By now you have noticed that all three moieties involve a five-membered ring associated with two oxygens. Because their mode of action is the inhibition of insecticide-metabolizing enzymes, it is likely that this steric three-dimensional structure is generally the most effective in enzyme binding.

The synergists are usually used in sprays prepared for the home and garden, stored grain, and on livestock, particularly in dairy barns. Synergists and the insecticides they synergize, such as pyrethrins, are quite expensive, thus seldom if ever used on crops.

First discovered in sesame oil, a material containing the methylenedioxyphenyl group was given the name *sesamin*. As mentioned earlier, many compounds having this moiety are synergistic, but only the structures of piperonyl butoxide, Sesamex, and Sulfoxide are shown.

PIPERONYL BUTOXIDE

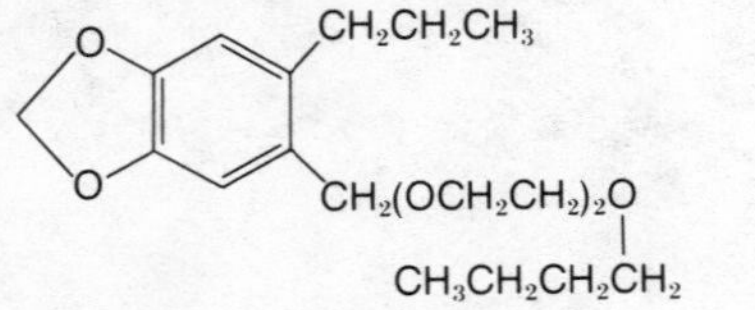

α-[2-(2-butoxyethoxy)ethoxy]-4,5-methylenedioxy-2-propyltoluene

SESAMEX

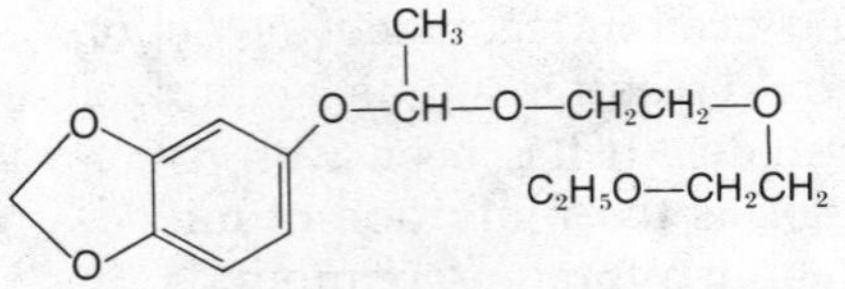

2-(2-ethoxyethoxy)ethyl-3,4-(methylenedioxy) phenyl acetal of acetaldehyde

SULFOXIDE

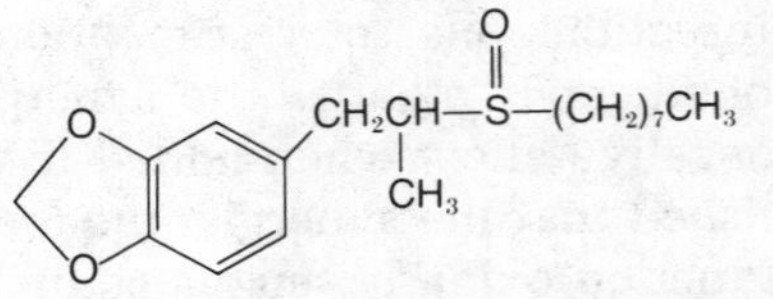

1,2-methylenedioxy-4-[2-(octylsulfinyl)propyl] benzene

Valone® and MGK 264 belong to the "no-name" moiety. Valone was discovered in 1942, whereas MGK 264 appeared in 1944. The latter has been used in the greatest quantity by far and has been used mostly in livestock and animal shelter sprays.

VALONE®

2-isovaleryl-1,3-indandione

MGK 264®

N-(2-ethylhexyl)-5-norbornene-2,3-dicarboximide

In summary, the synergists are used in practically all insecticide mixtures for the home, garden, and barn. Their mode of action is the temporary binding of enzymes that would otherwise metabolize the insecticide.

BOTANICALS

Botanical insecticides are of great interest to many, because they are "natural" insecticides, toxicants derived from plants. Historically, the plant materials have been in use longer than any other group, with the possible exception of sulfur. Tobacco, pyrethrin, derris, hellebore, quassia, camphor, and turpentine were some of the more important plant products in use before the organized search for insecticides had begun.

Some of the most widely used insecticides have come from plants. The flowers, leaves, and roots have been finely ground and used in this form, or the toxic ingredients have been extracted and used alone or in mixtures with other toxicants. Of the botanicals, only nicotine, pyrethrin, and rotenone will be discussed in this manual. Sabadilla and ryania were of commercial importance until about 1950.

Smoking tobacco was introduced to England in 1585 by Sir Walter Raleigh. As early as 1690, water extracts of tobacco were reported as being used to kill sucking insects on garden plants. As early as about 1890, the active principle in tobacco extracts was known to be nicotine, and, from that time on, extracts were sold as commercial insecticides for home, farm, and orchard. Today organic gardeners may soak a cigar or two in water overnight and spray insect-infested plants with the extract, achieving some success. "Black Leaf 40," which has long been a favorite garden spray, is a concentrate containing 40 percent nicotine sulfate. Today, nicotine is commercially extracted from tobacco by steam distillation or solvent extraction.

Nicotine is an alkaloid; it is a heterocyclic compound containing nitrogen and having prominent physiological properties. Other well-known alkaloids, which are not insecticides, are caffeine (found in tea and coffee), quinine (from cinchona bark), morphine (from the opium poppy), cocaine (from coca leaves), ricinine (a poison in castor oil beans), strychnine (from *Strychnos nux vomica*), coniine (from spotted hemlock, the poison that killed Socrates), and, finally, LSD (from the ergot fungus attacking grain), one of the banes of our twentieth-century culture.

As its mode of action, nicotine mimics acetylcholine (ACh) at the neuromuscular (nerve-muscle) junction in mammals, and results in

NICOTINE

1-3-(1-methyl-2-pyrrolidyl) pyridine

twitching, convulsions, and death, all in rapid order. In insects, the same action is observed, but only in the ganglia of their central nervous systems.

Flower heads of chrysanthemum, in the daisy family, are the source of pyrethrins. Pyrethrum powder was first used as an insecticide in the Transcaucasus region of Asia about 1800. Originally, the flower heads were ground and used in that form for louse control in the Napoleonic Wars. Kenya, Africa, is now the primary source of pyrethrins, which are now extracted from the flower heads with solvents.

Sprays containing pyrethrin are ideal home insecticides because of their extremely rapid knockdown (the instant action much favored by the American public), and they are probably the safest insecticides known for humans and their domestic animals. When pyrethrins are used alone, downed insects may recover and return to annoy again. Consequently, the synergists mentioned in the last section were discovered for the sole purpose of preventing insect survivors. The synergist prevents the insect from degrading pyrethrin and recovering.

Pyrethrins are a mixture of four compounds: pyrethrins I and II and cinerin I and II. Their structures can be assembled by attaching the R_1 and R_2 in their proper positions on the large ester structure to the left.

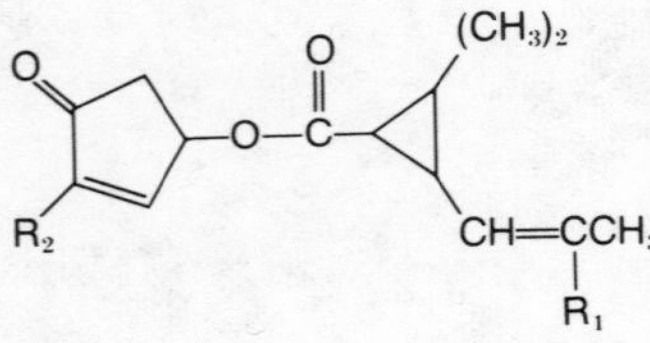

PYRETHRIN I

$R_1 = -CH_3$
$R_2 = -CH_2CH=CHCH=CH_2$

CINERIN I

$R_1 = -CH_3$
$R_2 = -CH_2CH=CHCH_3$

PYRETHRIN II

$R_1 = -\underset{\underset{O}{\|}}{C}-OCH_3$

$R_2 = -CH_2CH=CHCH=CH_2$

CINERIN II

$R_1 = -\underset{\underset{O}{\|}}{C}-OCH_3$

$R_2 = -CH_2CH=CHCH_3$

Despite its long use and study, the pyrethrin mechanism of action is still a mystery. The quick knockdown of flying insects is the result of rapid paralysis, which could be the effect on either muscle or nerves.

Rotenoids, the rotenone-related materials, have been used as crop insecticides since 1848, when they were applied to plants to control leaf-eating caterpillars. However, they have been used for centuries (at least since 1649) in South America to paralyze fish, causing them to surface.

Rotenoids are produced in the roots of two genera of the legume (bean) family, *Derris,* grown in Malaya and the East Indies, and *Lonchocarpus* (also names called *cubeb* or *cubé*), grown in South America.

Rotenone was apparently a totally safe garden insecticide, resulting in its popularity over the last 50 years. However, its registrations were cancelled by the EPA due to questionable physiological effects observed in laboratory animals fed rotenone over an extended period. It is highly toxic to most insects, and its mode of action is the interference in energy production through phosphorylation of adenosine diphosphate (ADP) to adenosine triphosphate (ATP). Such interference blocks oxidative phosphorylation. We are at a loss to explain its extreme toxicity to fish and insects and its low toxicity to mammals.

Rotenone is the most useful piscicide available for reclaiming lakes for game fishing. It eliminates all fish, closing the lake to reintroduction of rough species. After treatment, the lake can be restocked with the desired species. Rotenone is a selective piscicide in that it kills all fish at dosages that are relatively nontoxic to fish food organisms. It also breaks down quickly, leaving no residues harmful to the fish used for restocking. The recommended rate is 0.5 part of rotenone to one million parts of water (ppm), or 1.36 pounds per acre-foot of water.

Botanical insecticide use reached its maximum in the United States in 1966 and has declined steadily since. Pyrethrin is now the only botanical of significance in use, and this is typically in rapid knockdown sprays in combination with synergists and one or more synthetic organic insecticides formulated for use in the home and garden.

ROTENONE

1,2,12,12a,tetrahydro-2-isopropenyl-8,9-dimethoxy-[1]benzopyrano-[3,4-b]furo[2,3-b][1]benzopyran-6(6aH)one

Synthetic Pyrethroids

The natural insecticide pyrethrin has never been used for agricultural purposes because of its cost and instability in sunlight. Recently, however, several synthetic pyrethrin-like materials have become available only to agriculture and are referred to as synthetic pyrethroids. These materials are very stable in sunlight and are generally effective against most agricultural pests when used at the low rate of 0.1 lb per acre. These are permethrin (Ambush®, or Pounce®) and fenvalarate (Pydrin®).

FENVALARATE (Pydrin®)

Cyano (3-phenoxyphenyl) methyl 4-chloro-α-(1-methyl-ethyl) benzeneacetate

PERMETHRIN (Ambush®, Pounce®)

m-phenoxybenzyl (±)-cis, trans-3-(2,2-dichlorovinyl)-2,2-dimethylcyclopropanecarboxylate

THE INORGANICS

This group, too, is recorded for its historical significance. Inorganic insecticides are those that do not contain carbon. Usually they are white and crystalline, resembling the salts. They are stable chemicals, do not evaporate, and are frequently soluble in water.

Several inorganic materials have been used as insecticides. These include compounds of mercury, boron, thallium, arsenic, antimony, selenium, and fluoride. The only one of these used extensively today is arsenic, which is used in two forms, the arsenites (salts of arsenious acid) and the arsenates (salts of arsenic acid).

Paris green (green because of its copper content) was the first commonly used arsenical, a water-soluble arsenite. Next was lead arsenate. Finally, the third and last of these arsenicals was calcium arsenate, which was used for a time on vegetables in the 1930s and on cotton in the 1930s and 1940s.

Arsenicals are truly stomach poisons, exerting their toxic action following ingestion by the insects. Their action is attributed to the arsenite or arsenate ion, as will be explained later.

The arsenicals have a rather complex mode of action. First, they uncouple oxidative phosphorylation (by substitution of the arsenite ion for phosphorus), a major energy-producing step of the cell. Second, the arsenate ion inhibits certain enzymes that contain sulfhydryl (—SH) groups. And, finally, both the arsenite and arsenate ions coagulate protein by causing the shape or configuration of proteins to change.

Arsenical insecticides were very useful agricultural tools from 1930 until 1956, as we were making the transition from the simple to the complex synthetic molecules. They were, in fact, responsible for the initiation of large-scale pesticide applications eventually leading to the intensive use of fungicides and herbicides in modern agriculture.

Fluorine insecticides also included organic fluorine compounds, but these were of little importance and seldom used. The inorganic fluorides were sodium fluoride, used for cockroach and ant control around the home, and barium fluosilicate, sodium silicofluoride, and cryolite (NaF, $BaSiF_6$, and Na_3AlF_6, respectively). The last three were used for a time in plant protection. The fluoride ion inhibits many enzymes that contain iron, calcium, and magnesium. Several of these enzymes are involved in energy production in cells, as in the case of phosphatases and phosphorylases. A similar but indirect action can be assigned to the fluoroacetate ion ($FCH_2\overset{\displaystyle O}{\overset{\|}{C}}O—$), although it is an organofluorine.

The last group of inorganics are the silica gels or silica aerogels. These are light, white, fluffy silicates used for household insect control. The silica aerogels kill insects by adsorbing waxes from the insect cuticle permitting the continuous loss of water from the insect body. The insects then become desiccated and die from dehydration.

FUMIGANTS

The fumigants are small, volatile, organic molecules that become gases at temperatures above 40°F. They are usually heavier than air and commonly contain one or more of the halogens (Cl, Br, or F). Most are highly penetrating, reaching through large masses of material. They are used to kill insects, insect eggs, and certain microorganisms in buildings, warehouses, grain elevators, soils, and greenhouses and in packaged products such as dried fruits, beans, grain, and breakfast cereals.

Fumigants, as a group, are narcotics. That is, their mode of action is more physical than chemical. The fumigants are liposoluble (fat soluble); they have common symptomology; their effects are reversible; and their activity is altered very little by structural changes in their molecules. Narcotics induce narcosis, sleep, or unconsciousness, which in effect is their action on insects.

Liposolubility appears to be an important factor in the action of fumigants, since these narcotics lodge in lipid-containing tissues, which are found in the nervous system. There are a few of the fumigants whose mode of action is more than narcotic. These are ethylene dibromide, hydrogen cyanide, and chloropicrin. Because of its strong "tear-gas" effect on humans, chloropicrin is commonly added in trace quantities to many fumigants as an olfactory indicator or warning gas. Otherwise many of the fumigants would be lethal to humans before they became aware of a dangerous concentration in the air breathed.

Some of the more common fumigants' names and structures are presented in the list to the right.

Methyl bromide	CH_3Br
Ethylene dibromide	$BrCH_2CH_2Br$
Ethylene dichloride	$ClCH_2CH_2Cl$
Hydrogen cyanide	HCN
Chloropicrin	Cl_3CNO_2
Sulfuryl fluoride (Vikane®)	SO_2F_2
Vapam	$CH_3NHC(=S)-S-Na$
Telone	$ClCH=CH-CH_2Cl$
D-D®	$CHCl=CHCH_2Cl$ plus $ClCH_2CHCH_3$ with Cl
Chlorothene	CH_3CCl_3
Nemagon® (DCBP)	$BrCH_2CHCH_2Cl$ with Br
Ethylene oxide	CH_2-CH_2 with O
Naphthalene (crystals)	
p-dichlorobenzene (PDB crystals)	

MICROBIALS

Microbial insecticides obtain their name from microbes or microorganisms that are used to control certain insects. Like mammals, insects also have diseases caused by fungi, bacteria, and viruses. In several instances, these have been isolated, cultured, and mass-produced for use as pesticides.

The insect disease-causing microorganisms do not harm other animals or plants. The reverse of this is also true. This method of insect control is ideal in that the diseases are usually rather specific. Undoubtedly the future holds many such materials in the arsenal of insecticides, since several new insect pathogens are identified each year. However, at the present only a few are produced commercially and approved by the Environmental Protection Agency for use on food and feed crops. There is still some concern regarding the very remote chance of human susceptibility to these diseases, hence the slow advances into this relatively new field and exceptional precautionary testing.

The microorganism *Bacillus thuringiensis* is a disease-causing bacterium whose spores are necessary for disease induction. These spores produce compounds that injure the gut of insect larvae in such a way that invasion of the body cavity follows. This organism produces four substances toxic to insects: (1) a crystalline protein that paralyzes the gut of caterpillars; (2) a toxin that passes through the gut and kills certain fly maggots and pupae; (3) phospholipase C, an enzyme that breaks down the insect cell wall; and (4) another phospholipase that acts on phospholipids. The active agents of *Bacillus thuringiensis* are large protein molecules whose structures are not known.

Several insect viruses are in the development stage, all of which are aimed at caterpillars and which are only experimentally available. Viruses are highly specific and have modes of action that may not be identical throughout. Generally, the viruses result in crystalline proteins that are eaten by the larva and begin their activity in the gut. The virus unit then passes through the gut wall and into the blood. From there, several possibilities occur, but the known aspects are that the units multiply rapidly and take over complete genetic control of the cells, causing their death.

In summary, the microbial insecticides offer great promise for the future. Several are available for agricultural use, while others are in the experimental or development stage. *Bacillus thuringiensis* has become available as Thuricide® and Dipel® for garden and ornamental use by homeowners and will control most caterpillars. The average suburbanite may not become particularly fond of them because they require several days to kill, unlike the speedy effects we have come to expect from the traditional insecticides.

INSECT RESISTANCE TO INSECTICIDES

Before closing this chapter on insecticides, the continually growing problem of resistance should be discussed. The terms *resistance* and *tolerance* are in fairly common use: Most staphylococcus or staph infections are resistant to penicillin; flies have developed resistance to DDT and other insecticides; cockroaches have become resistant to chlordane; some strains of the venereal disease, gonorrhea, cannot be controlled with any but the very latest antibiotics; and some recently developed fungicides with specific modes of toxic action fail to control certain plant diseases.

Resistance to insecticides by insects dramatically demonstrates a miniform of evolution. The susceptible insects are killed, leaving behind only those that are genetically resistant to the toxicant. Resistant individuals make up an increasingly large part of the pest population and pass their resistance on to the next and future generations. Resistance in insect pests simply represents the survivors of a stringent biological selection mechanism, the insecticide, over several generations. The greater the number of generations exposed to an insecticide, the greater the potential for developing resistance as a consequence of this intensive selection mechanism.

Resistance to chemicals by insects is a problem worthy of genuine concern. In 1944, only 44 insect species were known to have developed resistance to insecticides. Keep in mind that the new synthetic insecticides were not yet on the scene. Now, a third of a century later, estimates place this number beyond 250, more than half of which are agricultural pests.

As more insecticides become ineffective against certain species, the problems of insect pest control increase proportionally. Populations of several insect pests now have such a high proportion of individuals resistant to all known insecticides that substitute materials are no longer available and insecticides are not recommended as control measures. Chemical control in these instances is no longer the method of choice. We are literally exhausting our arsenal of chemical tools. Resistance to insecticides is not unique to pests of agriculture. Public health problems resulting from resistance in insects that transmit or vector disease may become far more important than our agricultural difficulties.

One of the not-so-well-hidden, but unfamiliar phases of insecticide resistance is cross-resistance. Insects that have developed resistance to one organochlorine insecticide are frequently resistant to another organochlorine, even though they have not been previously exposed. In many instances, insects that became resistant to the organochlorines quickly became resistant to the organophosphate insecticides. And because the modes of action of the OPs and carbamates are similar, insects that became resistant to the OPs soon became resistant to one or more of the carbamates.

These examples of resistance can be attributed to the singular mode of action of a particular insecticide, which disrupts only one genetically controlled process in the metabolism of the insect. The result is that resistant populations appear suddenly, either by selection of resistant individuals in a population or by a mutation, which appears to be the less common of the two routes. Generally, the more specific the site and mode of insecticidal action, the greater the likelihood that an insect will develop resistance to that chemical.

The seeming ineffectiveness of an insecticide does not indicate with certainty that the insect is resistant. Effectiveness can also be reduced by the destruction of natural controls; for example, of parasites and predators that normally held the pest species in check. Only a very small fraction of the total number of insects are considered pests. At normal population densities, most insects pose no threat to cultivated crops, and many are important to the health and stability of the environment because they control other potentially damaging species.

Most of the insecticides in use today have broad-spectrum effects, that is, they are lethal to a wide range of insects, including beneficial as well as pest species. When insect pest populations are drastically reduced, their natural enemies are reduced even more. A resurgence in the pest population can then occur, resulting in increased damage to the crop presumably being protected.

With the use of these broad-spectrum insecticides, insects that were controlled naturally are sometimes caused to increase to such numbers that they become pests. This is the result of killing the insect's natural enemies. One quickly sees that an insect can be made a pest by "proper" use of insecticides applied to control a primary pest. In this context, this "proper" use could become "improper."

Resistance is indeed a problem in many insect pest species, and the problem is not going to go away or resolve itself. The insecticides to which an insect species has become resistant are no longer useful tools, compelling us to find new tools with new and different modes of action. Unfortunately, the exploration and discovery of new insecticides has not kept pace with their loss at the resistance end, coupled with the legal removal of others for reasons of human and environmental safety. Thus, the long-range answers to insect resistance will have to be found in other control methods or in the combination of methods known as *integrated pest management*. This subject is discussed in more detail under the final section of Chapter 22, "Alternatives to Pesticides: Where Do We Go from Here?"

CHAPTER 7

Molluscicides

Four and twenty tailors went to
  kill a snail,
The best man among them durst not
  touch her tail.
She put out her horns like a little
  Kyloe cow,
Run, tailors, run, or she'll kill
  you all e'en now.

Old English Nursery Rhyme

Molluscicides are compounds used to control snails, which are the intermediate hosts of parasites of medical importance to humans and which feed in gardens, greenhouses, and fields, including the slugs. Medically, snails are extremely important, especially freshwater snails, such as those which serve as intermediate hosts of organisms causing schistosomiasis and fascioliasis in humans, and lung and liver fluke intermediates of humans, dogs, cats, and domestic animals. Economically, the giant African snail, *Achatina fulica,* is considered to be the most important land snail pest known. This particular snail was introduced into Miami, Florida, in 1966, by an 8-year-old boy who returned from a vacation in Hawaii with three of these cute pests in his pockets.

The molluscicide known as *metaldehyde* has been used as baits for the control of slugs and snails commercially and around the home since its discovery in 1936. Its continued use and success can be attributed both to its attractant and toxicant qualities. Other materials used in various formulations, e.g., baits, sprays, fumigants, and contact toxicants, include metaldehyde plus calcium arsenate, sodium arsenite, ashes, copper sulfate, carbon disulfide, chlordane, coal tar, DDT, lindane, hydrogen cyanide, kerosene emulsion, methyl bromide gas, sodium chloride, and sodium dinitroorthocresylate.

Clonitralid (Bayluscide®) is one of the most promising molluscicides discovered. It is especially toxic to freshwater snails that serve as intermediate hosts of the organisms causing schistosomiasis and fascioliasis, two of the more tragic diseases of human beings. It is also useful as a piscicide but has undesirable toxic side effects on other aquatic organisms, such as crayfish, frogs, and clams, with little effect on plankton and aquatic vegetation.

METALDEHYDE

polymer of acetaldehyde, or metacetaldehyde

CLONITRALID (Bayluscide®)

2′,5-dichloro-4′-nitrosalicylanilide,
2-aminoethanol salt

FENTIN ACETATE (Brestan®)

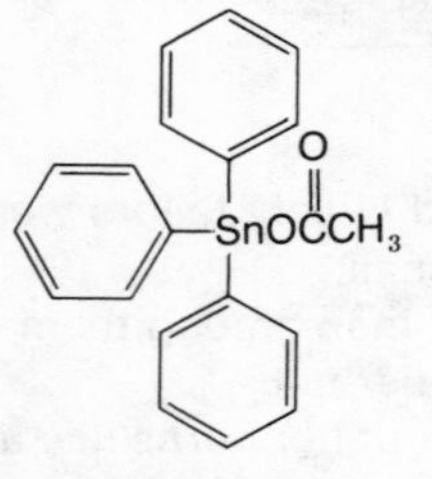

triphenyltin acetate

FORMETANATE (Carzol®)

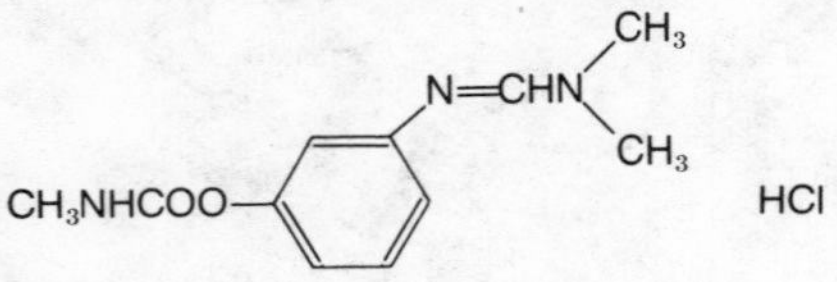

CH_3NHCOO HCl

(3-dimethylamino-(methylene-iminophenyl))-
N-methylcarbamate

TRIFENMORPH (Frescon®)

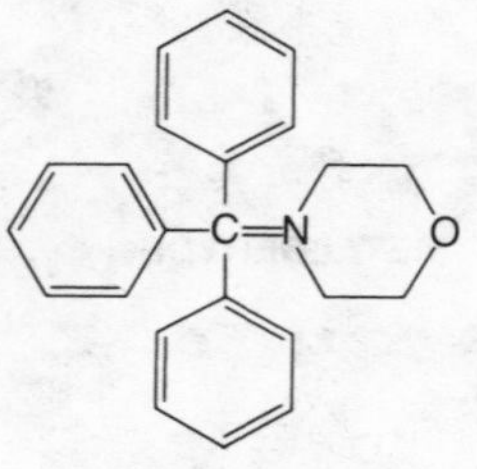

N-tritylmorpholine

Triphenyltin acetate is also used as a fungicide and algicide in addition to its molluscicide properties.

Formetanate (Carzol®), although not currently registered by EPA as a molluscicide, shows great promise as a garden snail and slug bait. It is registered as an acaricide-insecticide.

Trifenmorph (Frescon®) has been used very successfully to control aquatic snails that transmit bilharzia in humans, and the aquatic and semiaquatic snails that are the intermediate hosts of the organisms causing fascioliasis.

PCP, or pentachlorophenol, which we met earlier as both a fungicide and a herbicide, is one of the phenolic compounds that is universally toxic. Thus, it is applied as a molluscicide to control snail carriers of larval human blood flukes causing schistosomiasis in Egypt. PCP is available as a formulated product to be applied with petroleum solvent or as an emulsion.

Polystream is a mixture of chlorinated benzenes proved effective against oyster drill, a predatory snail. It is both molluscicidal and repellent and is used as a selective molluscicide on oyster grounds in Connecticut and New York waters.

Mercaptodimethur (Mesurol®) is another successful insecticide registered for use against snails and slugs on ornamentals. It is highly effective against these pests and has also demonstrated repellency to several bird species.

PCP

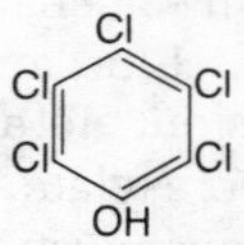

Pentachlorophenol

POLYSTREAM

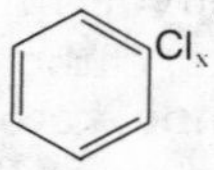

Cl_x

MERCAPTODIMETHUR (Mesurol®)

CH_3S

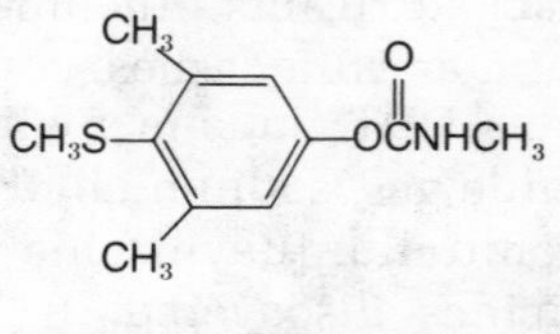

3,5-dimethyl-4-(methylthio)phenol
methyl carbamate

CHAPTER 8

Nematicides

O Rose, thou are sick!
The invisible worm
That flied in the night,
In the howling storm,

Has found out thy bed
Of crimson joy,
And his dark secret love
Does thy life destroy.

William Blake

The microscopic roundworms that live in soil or water are known as *nematodes*. Many are free-living, whereas others are parasitic on plants or animals. Some species of nematodes inadvertently introduce pathogenic root-invading microorganisms into the plants while feeding. Nematodes may also predispose crop-plant varieties to other disease-causing agents, such as wilts and root rots. In other instances, the nematodes themselves cause the disease, disrupting the flow of water and nutrients in the xylem system, resulting in root-knot or deprivation of the above-ground parts, and ultimately causing stunting.

Nematodes are covered with an impermeable cuticle, which provides them with considerable protection. Chemicals with outstanding penetration characteristics are therefore required for their control.

Nematicides are seldom used by homeowners except in a greenhouse or cold frames. For the most part, we can say that nematicides are not and should not be used by the layperson, mainly because of their hazard. Those that are available commercially fall into four groups: (1) halogenated hydrocarbons (some of which were described earlier); (2) organophosphate insecticides; (3) isothiocyanates; and (4) carbamate or oxime insecticides.

Most of today's nematicides are soil fumigants, volatile halogenated hydrocarbons. To be successful, they must have a high vapor pressure, to spead through the soil and to contact nematodes in the water films surrounding soil particles.

HALOGENATED HYDROCARBONS

The nematicidal properties of DD and EDB were discovered in 1943 and 1945 and effectively launched the use of volatile nematicides on a field-scale basis. Previously only seedbeds, greenhouse beds, and potting soil had been treated, with materials such as chloropicrin, carbon disulfide, and formaldehyde. These were very expensive, in some instances explosive, and usually required a surface seal, because of their relatively high vapor pressures.

DICHLOROPROPENE-DICHLOROPROPANE (D-D®)

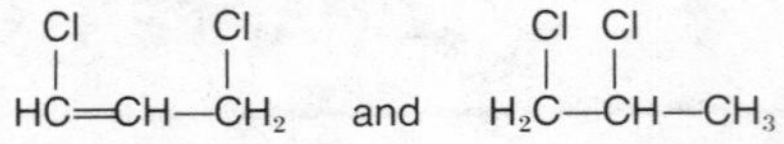

$$
\underset{\text{1,3-dichloropropene}}{HC=CH-CH_2} \quad \text{and} \quad \underset{\text{1,2-dichloropropane}}{H_2C-CH-CH_3}
$$

with Cl substituents, and

EDB

$$Br-CH_2-CH_2-Br$$

ethylenedibromide

DD and EDB are both injected into the soil several days before planting, to kill nematodes, eggs, and soil insects.

DBCP is one of the easier-to-use nematicides, in that it is formulated as both the emulsifiable concentrate or as granules. It can be used as postplant treatments and is registered for use on numerous fruit and vegetable crops. This is the only material in its class that can be used near living plants. In August 1977, it was announced by its manufacturers that DBCP was found to cause sterility in male workers involved in its manufacture. Subsequently it has been removed from the market and is currently available only as a restricted use pesticide for a very few essential needs. Because of this health hazard DBCP is no longer available to the homeowner.

DBCP

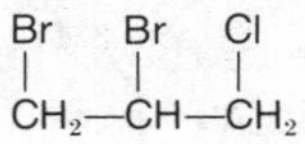

$$
\underset{\text{dibromochloropropane}}{CH_2-CH-CH_2}
$$

with Br, Br, Cl substituents

Methyl bromide is a first-class, all-around fumigant, lethal to all plant and animal life, hence its classification as a sterilant. Since it is totally phytotoxic, it must be used as a preplant application, followed by adequate aeration time. A two-week waiting period after fumigation is an acceptable rule-of-thumb.

METHYL BROMIDE

$$CH_3Br$$

bromomethane

Tetrachlorothiophene is used as a preplant treatment of tobacco beds for nematode control; it is soluble in both water and petroleum solvents.

TETRACHLOROTHIOPHENE (Penphene®)

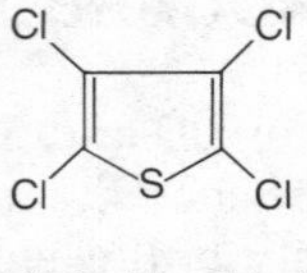

2,3,4,5-tetrachlorothiophene

Most of these halogenated hydrocarbon nematicides' mode of action is that of a narcotic fumigant. They are liposoluble and, as such, lodge in the primitive nervous systems of nematodes and kill primarily through physical rather than chemical action.

ORGANOPHOSPHATES

Nematodes have nervous systems similar to that of insects, although they are more primitive, and are thus susceptible to the action of the organophosphate insecticides. Unfortunately, most of the organophosphates commonly used as insecticides are degraded rapidly in the soil, and only the systemics are effective as nematicides. As a result, only a few of the insecticidal materials are involved, several of which have been developed especially for their nematicidal effects. All of these organophosphates inhibit the nerve-transmitter enzyme, cholinesterase, resulting in paralysis and, ultimately, the death of affected nematodes.

DICHLOFENTHION (Nemacide®)

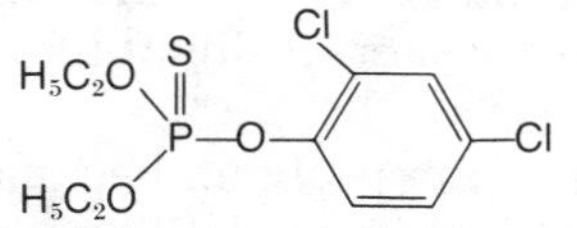

diethyl dichlorophenyl phosphorothioate

DISULFOTON (Disyston®)

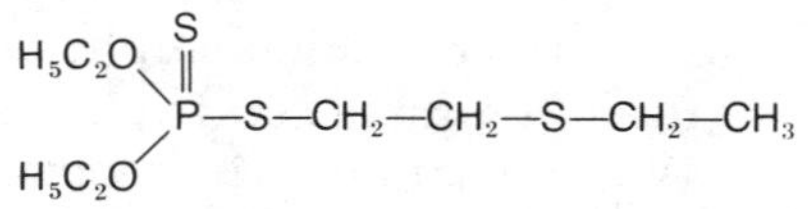

O,O-diethyl S(2-(ethylthio)ethyl phosphorodithioate

PROPHOS (Mocap®)

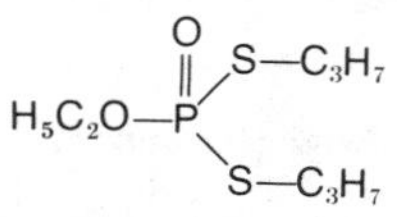

O-ethyl S;S-dipropyl phosphorodithioate

FENSULFOTHION (Dasanit®)

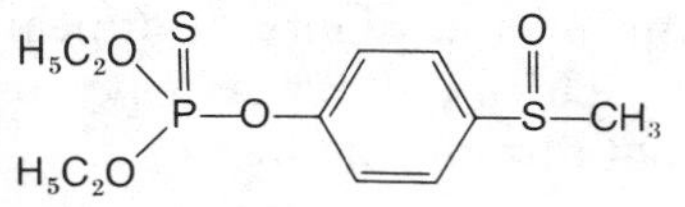

O,O-diethyl-O-[p-(methylsulfinyl)phenyl] phosphorothioate

PHORATE (Thimet®)

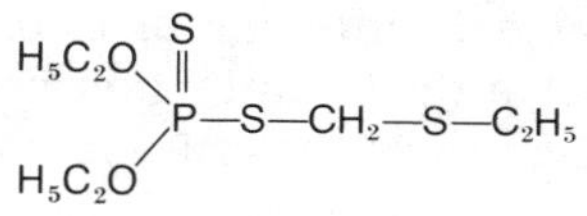

O,O-diethyl S-(ethylthio)methyl phosphorodithioate

THIONAZIN (Nemophos®, Zinophos®)

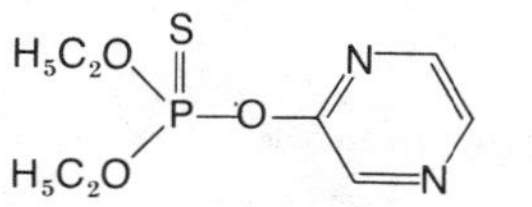

O,O-diethyl,O-(2-pyrazinyl) phosphorothionate

ISOTHIOCYANATES

There are three nematicides in this classification: SMDC, MIT, and DMTT. SMDC is a dithiocarbamate mentioned under the fungicides, but it is readily converted to an isothiocyanate, as we have learned, and is active against all living matter in the soil.

MIT, also discussed with the fungicides, is a preplant fumigant effective against nematodes, fungi, and weeds.

DMTT is a diazine, resembling the thiazoles slightly and undergoing a similar ring cleavage in the soil to produce methylisothiocyanate.

DMTT

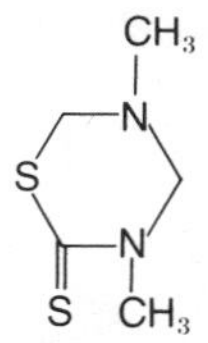

tetrahydro-3,5-dimethyl-2H,1,3,5-thiadiazine-2-thione

SMDC (Vapam®)

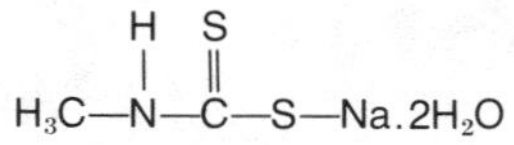

sodium N-methyldithiocarbamate dihydrate

MIT (Vorlex®)

$CH_3—N≡C≡S$

methylisothiocyanate

ALDICARB (Temik®)

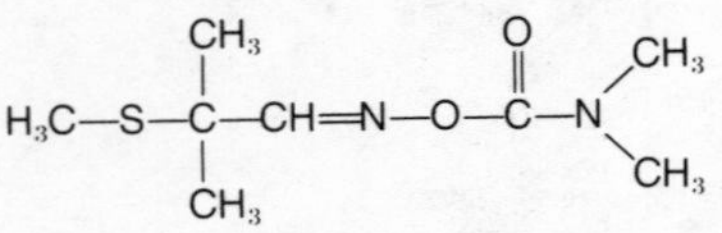

2-methyl-2-(methylthio)propionaldehyde
O-(methylcarbamoyl) oxime

CARBOFURAN (Furadan®)

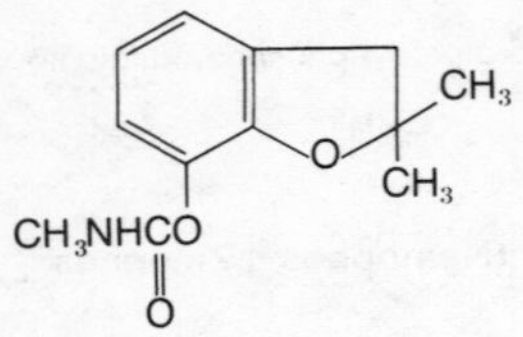

2,3-dihydro-2,2-dimethyl-7-benzofuranyl
methylcarbamate

CARBAMATES

The carbamate aldicarb is the only example of an oxime nematicide, and was mentioned earlier, in the section on carbamate insecticides as a systemic. Although registered for use but on a few crops at present, it shows great promise both as insecticide and nematicide.

Unlike most of the other nematicides, aldicarb is a solid and is formulated as the granular material. It is drilled into the soil at planting, or after the plants are in various stages of growth. It becomes water soluble and is absorbed by the roots and translocated throughout the plant.

Carbofuran is a very recent and promising systemic carbamate insecticide-nematicide. It has a relatively short residual life, making it useful on forage and vegetable crops.

With nematicides in general, the residue problem becomes more serious, given the increased use of stable, nonvolatile compounds such as the carbamates and organophosphates. Their long residual effect may limit the use of some compounds despite their high toxicity to nematodes. Residues in plants grown in treated soil may be prohibited by the Environmental Protection Agency or may be unpalatable to humans or animals consuming the plants. It appears, then, that the volatile, nonresidual nematicides will be in heavy and continued use for years to come—but not around the home.

CHEMICALS USED IN
THE CONTROL OF VERTEBRATES:
ANIMALS WITH BACKBONES

CHAPTER 9

Rodenticides

Rats!
They fought the dogs and killed the cats,
  And bit the babies in the cradles,
And ate the cheeses out of the vats,
  And licked the soup from the cooks'
    own ladles.

Robert Browning, *The Pied Piper of Hamelin*

Several small mammals, especially rodents, damage human dwellings, stored products, and cultivated crops. Among these are native rats and mice, squirrels, woodchucks, pocket gophers, hares, and rabbits. Rats are notorious freeloaders, and, in some of the underprivileged countries, where it is necessary to store grain in the open, as much as 20 percent may be consumed by rats before people can do so.

The number of rodent species (order *Rodentia*) comprises about one-half of all mammalian species, and, because they are so very highly prolific and widespread, they are continuously competing with humans for food. Most of the methods used to control rodents are aimed at destroying them. Poisoning, shooting, trapping, and fumigation are among the methods selected. Of these, poisoning is most widely used and is probably the most effective and economical. Because rodent control is in itself a diverse and complicated subject, we will mention only those more commonly used rodenticides.

Rodenticides differ widely in their chemical nature. Strange to say, they also differ widely in the hazard they present under practical conditions, even though all of them are used to kill animals that are physiologically similar to humans.

PHOSPHORUS

Phosphorus occurs in two common forms, the relatively harmless red and the highly toxic white or yellow phosphorus. Phosphorus is seldom used today, having been replaced by the anticoagulants. Yellow phosphorus attacks the liver, kidney, and heart, resulting in tissue disintegration; it also causes rats to attempt to vomit, a function that, uniquely, they cannot perform.

Zinc phosphide (Zn_3P_2) is an intense poison to mammals and birds and is used against rats, mice, and gophers. Its mode of action is similar to that of phosphorus.

Gophacide® is a very successful organophosphate rodenticide developed in Germany, but it is little used in the United States. You will recall that the organophosphates are inhibitors of cholinesterase.

GOPHACIDE®

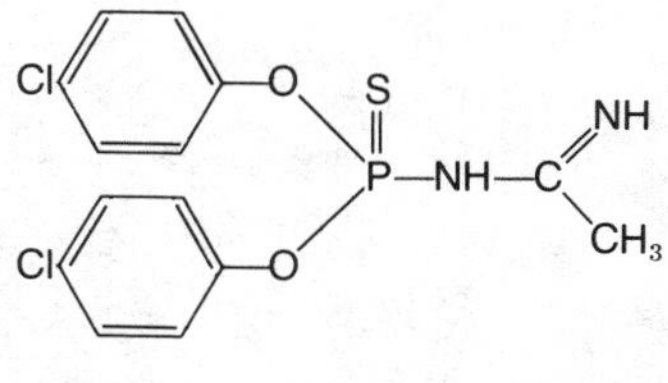

O,O-Bis(4-chlorophenyl)acetimidoyl-
phosphoramidothioate

DICUMAROL

3,3'-methylene bis (4-hydroxycoumarin)

WARFARIN

3-(α'-acetonylbenzyl)-4-hydroxycoumarin

COUMACHLOR

3-(α-Acetonyl-4-chlorobenzyl)-4-hydroxycoumarin

COUMARINS (ANTICOAGULANTS)

The most successful group of rodenticides are the coumarins, represented classically by Warfarin. There are five compounds belonging to this classification, all of which have been very successful rodenticides. Their mode of action is twofold: (1) inhibition of prothrombin formation, the material in blood responsible for clotting; and (2) capillary damage, resulting in internal bleeding. The coumarins require repeated ingestion over a period of several days, leaving the unsuspecting rodents growing weaker daily. The coumarins are thus considered relatively safe, since repeated accidental ingestion would be required to produce serious illness. In the case of most other rodenticides, a single accidental ingestion could be fatal. Of the several types of rodenticides available, only those with anticoagulant properties are safe to use around the home.

The first coumarin was Dicumarol, introduced in 1948, and was hit on after identifying the molecule as the compound responsible for sweet clover's toxicity to cattle. As a rodenticide, however, dicumarol has been superseded by Warfarin. Warfarin was released in 1950 by Wisconsin Alumni Research Foundation (thus its name *WARF coumarin*, or *WARFarin*). It was immediately successful as a rat poison, because rats did not develop "bait shyness," as they did with other baits during the required ingestion period of several days. Coumachlor was introduced in 1953, but has never been successful in the U.S. because of Warfarin's wide acceptance. Coumatetralyl was developed in Germany and introduced in the United States in 1957 with a fair degree of success in situations where Warfarin had resulted in bait shyness of rodents. Finally, Fumarin® is a commonly used material.

COUMATETRALYL

3-(d-tetralyl)-4-hydroxycoumarin

FUMARIN®

3-(α-acetonylfurfuryl)-4-hydroxycoumarin

ORGANOCHLORINES

DDT 50 percent dust was used for years by structural pest control operators as a tracking powder. The dust was sprinkled in the known runs of mice, and, after tracking through the dust, the mice stopped to preen themselves and clean their feet. Death resulted from convulsions and paralysis, just as in insects.

Similarly, endrin was sprayed on orchard soils and fruit tree trunks in rather heavy concentration during the fall or winter months. Field mice eating the bark of fruit trees and trailing across treated soil quickly ingested lethal quantities. Although quite effective, neither practice is acceptable today, because of persistence and hazard to other species. (See structures in section on organochlorine insecticides.)

BOTANICALS

Red Squill, which comes from the powdered bulbs of a plant, Mediterranean squill, was used before 1935, but was never more than a mediocre rodenticide. The active ingredient is an alkaloid, scillarenin, classed as a cardiac glycoside. Its specific activity is due to the inability of rats to vomit—thus they must absorb the toxicant. Other animals ingesting squill do vomit, which permits them to survive accidental poisoning.

Strychnine is an alkaloid, from an Asiatic tree, *Strychnos nux-vomica*, that is usually converted to strychnine sulfate for use as a rodenticide. Strychnine is highly toxic to all warm-blooded animals and acts by paralyzing specific muscles, resulting in cessation of breathing and heart action. It is only a fair rodenticide and has been replaced by the anticoagulants.

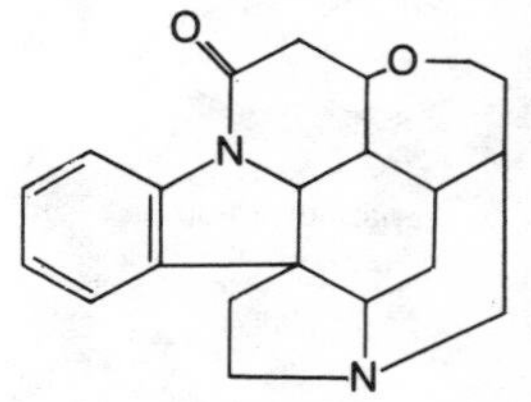

INDANDIONES

Three compounds, chlorophacinone, diphacinone, and pindone, belong to this class, have the same remarkable anticoagulant property as Warfarin, and have replaced Warfarin where rodent avoidance behavior (bait shyness) made it ineffective. Sold as baits containing up to 0.0075 percent of the toxicant, they must be ingested for several consecutive days before they become effective. Because of this characteristic, the anticoagulants provide a definite safety factor for children and animals.

RED SQUILL (Scillarenin)

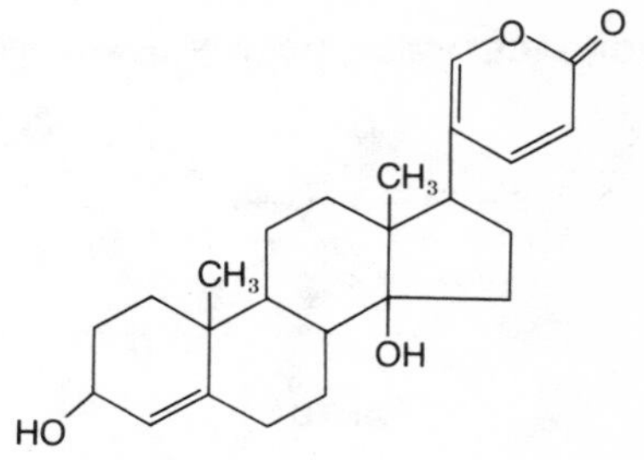

STRYCHNINE

CHLOROPHACINONE

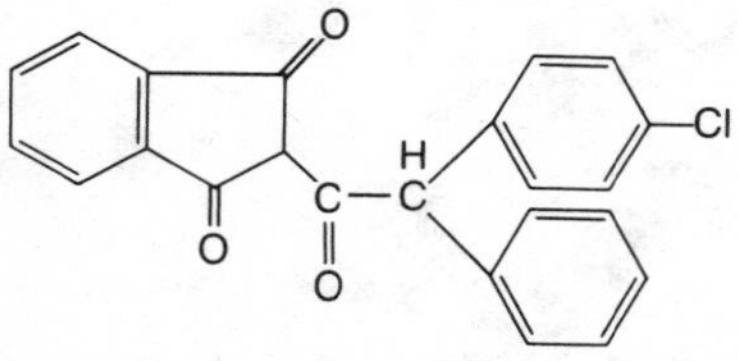

2-[(2-*p*-chlorophenyl)-2-phenylacetyl]-1,3-indandione

DIPHACINONE (Diphacin®)

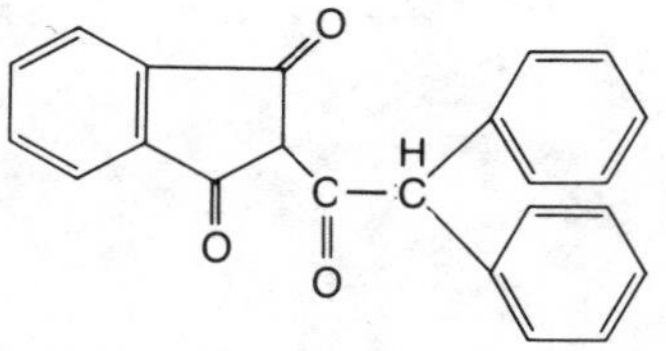

2-diphenylacetyl-1,3-indandione

PINDONE (Pival®)

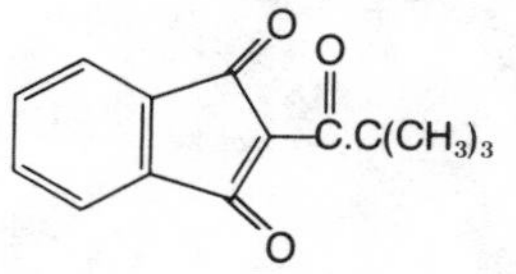

2-pivaloylindane-1,3-dione

SODIUM FLUOROACETATE (1080®)

$$F-CH_2-\overset{\overset{\displaystyle O}{\|}}{C}-O-Na$$

FLUOROACETAMIDE (1081, Fluorakil 100®)

$$F-CH_2-\overset{\overset{\displaystyle O}{\|}}{C}-O-NH_2$$

ANTU®

$$\overset{\overset{\displaystyle S}{\|}}{NHCNH_2}$$

α-naphthylthiourea

VACOR®

$$CH_2.NH-\overset{\overset{\displaystyle O}{\|}}{C}-NH-\!\!\!\!\!\!-NO_2$$

N-3-pyridylmethyl-N'-p-nitrophenyl urea

MISCELLANEOUS RODENTICIDES

Compound 1080, sodium fluoroacetate, one of the most toxic poisons known to warm-blooded animals, was introduced in 1947. Its use is now restricted to authorized trained personnel because of its extreme hazard to humans and domestic animals. Sodium fluoroacetate has a strong effect on both the heart and nervous system, resulting in convulsions, paralysis, and death. More recently it has made the news in coyote control programs in the West.

Compound 1081 is a moderately fast-acting rodenticide closely related to sodium fluoroacetate. It possesses a lower mammalian toxicity and a longer latent period before animals become distressed and stop feeding. Its use is less likely to lead to poison shyness because of sublethal dosing.

Antu® is a cryptic form of the chemical nomenclature given it when introduced in 1946. Because rodents quickly develop a tolerance to Antu, it has been displaced by other materials.

Thallium sulfate ($TlSO_4$) is an old and reliable rodenticide that has also been replaced by the safer anticoagulants, for the same general reasons as for the others. It is a general cellular toxin that resembles arsenic in its effects and attacks or inhibits enzymes other than those containing —SH groups.

VACOR®

The latest entry into the rodenticide field is Vacor®, an acute toxicant that kills in one feeding. It is effective against rats that have developed resistance to Warfarin, which kills over a period of several days by causing internal hemorrhaging. Vacor® kills in 4 to 8 hours after ingestion of a single dose of as little as 0.5 gram of the 2 percent bait. It is specific against the most common rodent pests, including the Norwegian and roof rats and the house mouse. Mode of action is the inhibition of niacinamide metabolism, and the rodents die from paralysis and pulmonary arrest.

CHAPTER 10

Avicides

We accept the fact that all birds can, in one way or another, at times be beneficial to humans. They provide enjoyment and wholesome recreation for most of us regardless of where we live. Despite the fact that wild bird populations are for the most part beneficial, there are occasions when individuals of certain species can seriously compete with human interests. When these situations occur, control measures are inevitable.

These beautiful winged creatures create pest problems singly or in small groups, but especially when in large aggregations. Most of the areas of conflict with humans are: (1) destruction of agricultural foodstuffs and predation; (2) contamination of foodstuffs or defacing of buildings with their feces; (3) transmission of diseases, directly and indirectly, to man, poultry and dairy animals; (4) hazards at airports and freeways; and (5) being a general nuisance or affecting human comfort.

Of the various methods of control, this book deals with repellents, sticky chemicals on ledges and roots, toxic baits, soporifics (stupefacients), and surfactants (feather-wetting agents).

Probably the most prominent of the avicides is the relatively new Avitrol® (4-aminopyridine), which is used as a repellent; the effects result from the distress calls made by affected birds. The material has a relatively low LD_{50} of 20 mg/kg, thus mortality in some individuals is inevitable (LD_{50} means a dose lethal to 50 percent of organisms tested). It is to be used only by licensed pest control operators for driving away flocks of nuisance or feed-consuming birds from feedlots, fields, airports, warehouse premises, public buildings, and grain-processing plants. Only a small number of birds need be affected to alarm the rest of the flock. The treated grain must be eaten to be effective. After one alarming exposure, birds will not return to treated areas.

Starlicide is a chlorinated compound used as a slow-acting avicide against starlings and blackbirds. It is not effective against house sparrows. Since the material kills slowly, requiring from 1 to 3 days, large numbers of dead birds do not appear in the treated area but rather die in flight or at their roosts. It is formulated as 0.1 and 1.0 percent baits and available only to pest control operators trained in bird control.

AVITROL®

4-aminopyridine

STARLICIDE®, DCR 1339

3-chloro-*p*-toludine hydrochloride

ENDRIN

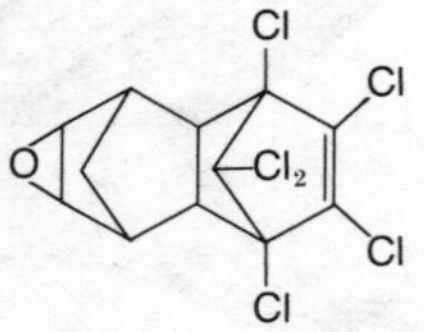

1,2,3,4,10,10-hexachloro-6,7-epoxy-1,4,4a,5,6,7,8,8a-
octahydro-1,4-*endo-endo*-5,8-dimethanonaphthalene

FENTHION (Baytex®)

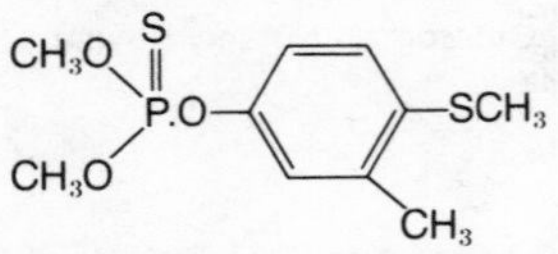

O,O-dimethyl *O*-[(4-methylthio)-*m*-tolyl]
phosphorothioate

The chlorinated cyclodiene insecticide Endrin has been used for years as an elevated, out-of-reach perch treatment for the control of pigeons, starlings, and English sparrows. It is quite effective, if used properly, and is available only to licensed pest control operators trained in bird control.

Another effective perch treatment is the organophosphate insecticide fenthion (Baytex®), used to control pigeons, starlings and English sparrows. This compound acts quickly after absorption through the feet of perching birds. Fenthion is for use only by licensed pest control operators and trained personnel of industry and government.

The old, general-purpose poison strychnine is registered by EPA as an avicide for the control of pigeons and English sparrows when used as a 0.6 percent grain bait. Strychnine acts very quickly, leaving the treated area strewn with dead birds, which should be removed at regular intervals. Prebaiting for several days is necessary before distributing the treated bait. For bird protection, treated grain should be colored before distribution, and uneaten bait should be removed.

Ornitrol® (SC-12937) is a new and apparently humane approach to bird control. It is a chemosterilant, a birth control agent for pigeons, designed to control population growth rather than to eradicate. It causes temporary sterility in pigeons after a 10-day feeding exposure by inhibiting egg production. It has little, if any, effect on mammals and is selective of pigeons when impregnated on whole corn grains too large for other species to feed on. It is formulated as a 0.1 percent bait.

STRYCHNINE

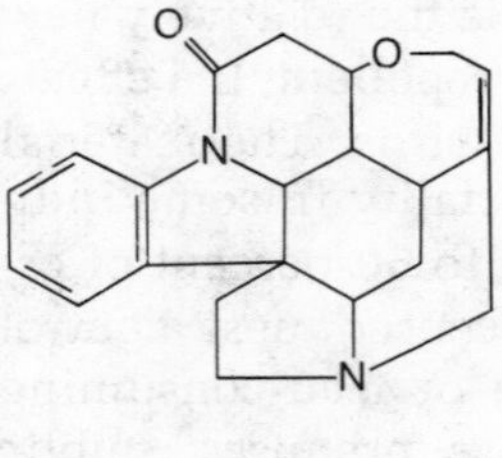

ORNITROL®, SC-12937

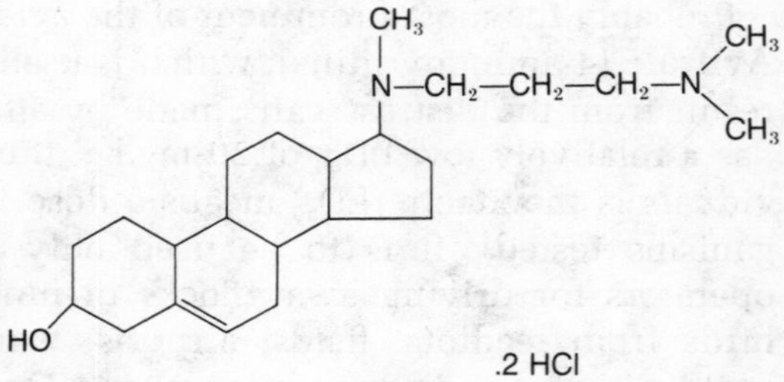

.2 HCl

20,25-diazacholesterol, hydrochloride

CHAPTER 11

Piscicides

Third Fisherman: Master, I marvel how the fishes live in the sea.

First Fisherman: Why, as men do aland; the great ones eat up the little ones.

William Shakespeare, *Pericles*

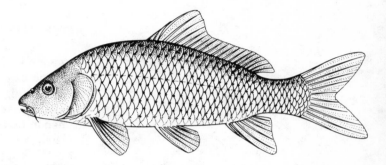

Piscicides are a small, heterogeneous assortment of chemicals that are rather nonspecific for fish but that are the best available. They are used to remove all fish from a body of water, for restocking with desirable game fish after some safe waiting interval.

Most pest fish conflicts involve undesirable species or rough or trash fish competing with more desirable game fish. There are also aesthetic problems: Fish may muddy the water or create odor problems as ponds and agricultural ditches dry up and are even known to choke municipal water intake pipes. The introduced species, such as minnows, carp, suckers, crappie, catfish, and bass, can on occasion create rough fish-sport fish conflicts, frequently requiring the poisoning of ponds, lakes, and streams by fisheries personnel and sportsmen, so that more desirable species can be established.

Pest fish species include the destructive sea lamprey, parasitic on commercial fish species, which gained access to the upper Great Lakes through the Welland Ship Canal built in 1829 to bypass Niagara Falls; the alewife, a native of the North Atlantic, which entered the Great Lakes in the same way as the sea lamprey; wild goldfish, members of the carp family, established as a result of the home-aquarist and bait-fish trades; the walking catfish, which escaped from a tropical fish dealer in Florida; and the carp, introduced into the United States more than 100 years ago from Germany. These are just a few examples of fish that, for one reason or another, must be removed by poisoning. Unfortunately, with the exception of the lamprey eel, the piscicides are nonspecific with regard to fish.

Less than 25 years ago, formal investigations were initiated to find and perfect control chemicals specific to fish. Previous to that, our fish poisons were borrowed from the agricultural insecticides and used without regard for the unique complexities of aquatic environments.

CHLORINATED INSECTICIDES

Many of the organochlorine insecticides were, early in their use history, found to be extremely toxic to fish and were in a few cases used as piscicides. Among these were toxaphene, beginning in 1948, Endrin, 1952, and endosulfan (Thiodan®) in 1960. Their use was usually catastrophic, since they are toxic to all forms of wildlife and very persistent. Consequently, none of the three were ever registered for use. (Endrin gained a bad reputation in the Mississippi River fish kills of the mid-1960s, as did endosulfan in the accidental fish kill in the Rhine River in 1969.) Needless to say, none of the organochlorine insecticides should be used as piscicides.

ROTENONE

ROTENONE

Rotenone is the most useful piscicide available for reclaiming lakes for game fishing. It eliminates all fish, closing the lake to reintroduction of rough species. After treatment, the lake can be restocked with the desired species. Rotenone is a selective piscicide, in that it kills all fish at dosages that are relatively nontoxic to fish food organisms. It also breaks down quickly, leaving no residues harmful to fish used for restocking. The recommended rate is 0.5 ppm or 1.36 pounds per acre-foot of water.

Rotenone is a strong inhibitor of the respiratory chain in fish, and the site of action is located in the flavoprotein region of this chain. The specialized structure of the gills favors passage of rotenone into the blood, which is then transported to vital organs for respiration inhibition.

ANTIMYCIN®

This is actually an antibiotic produced by Streptomyces fungi, whose fish poison characteristics were discovered in 1963. In the dosages used, it is specific for fish, leaving unharmed other aquatic life, waterfowl, and mammals. It is lethal in small concentrations to all stages of fish, egg through adult. It passes into the blood through the gills and irreversibly blocks cellular respiration at the cytochromes, thus inhibiting oxidative phosphorylation. Because it is not repellent to fish, it is the first piscicide to be successfully used at spot treatments in lakes. It degrades rapidly in water, usually within a few days, and even faster in high pH situations. Antimycin A, also sold under the name of Fintrol, is formulated as a 1 and 5 percent wettable powder and as a 10 percent solution, and must be used under technical supervision of state and federal fish and game agencies.

BAYLUSCIDE®

Bayluscide® (Clonitralid) is registered for the control of larval stages of the sea lamprey but is also outstanding against other fish and as a molluscicide. It is used primarily in combination with TFM (see next paragraph) to kill sea lamprey larvae in tributary streams of the Great Lakes. The liquid toxicant mixture is metered with precision into streams to kill lamprey residing in bottom muds, resulting in little harm to other aquatic organisms.

TFM

TFM (Lamprecide®) (3-trifluoromethyl-4-nitrophenol, sodium salt) is a selective toxicant against the larval stages of the sea lampreys, pests of the Great Lakes, and resulted from research by biologists of the U.S. Fish and Wildlife Service. The later discovery of synergistic action between it and Bayluscide® (see preceding paragraph) gave the Great Lakes Fishery Commission a highly effective piscicide for eliminating most of the larval stages of the lamprey harbored in hundreds of miles of tributary streams.

Despite the successes of reducing lamprey populations in Lake Superior especially, the lamprey continues to cause great losses to the Great Lakes fishing industry. In large bodies of water, even when the target pest is confined to tributary streams for breeding, the size of this breeding ground may be too large for chemical control.

CLONITRALID (Bayluscide®)

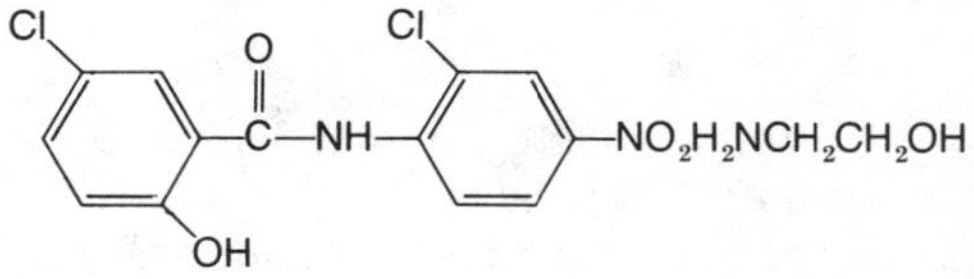

2′,5-dichloro-4′-nitrosalicylanilide,
2-aminoethanol salt

TFM (Lamprecide®)

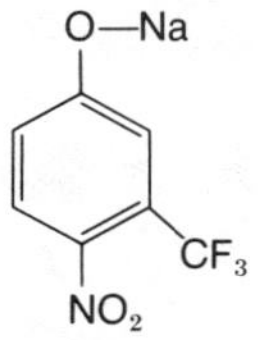

3-trifluoromethyl-4-nitrophenol, sodium salt

CHAPTER 12

Repellents

Do what we can, summer will have
its flies. If we walk in the
woods, we must feed mosquitoes.

Ralph Waldo Emerson, *Essays*

Most of us think of repellents strictly in the sense of driving away
mosquitoes, biting flies, and gnats—insect repellents. These are the
best known. There are, however, repellents for birds, dogs, deer,
moles, rabbits, and rodents. Have you ever wished for some way to
prevent the dog next door from defecating in your lawn or urinating
on your shrubs? Well, perhaps you may find a solution here.

INSECT REPELLENTS

Historically, repellents have included smokes, plants hung in
dwellings or rubbed on the skin as the fresh plant or brews, oils,
pitches, tars, and various earths applied to the body. Camel urine
sprinkled on the clothing has been of value, although questionable,
in certain locales. Camphor crystals sprinkled among woolens have
been used for decades to repel clothes moths. Before humans
developed a more edified approach to insect olefaction and behavior,
it was assumed that if a substance was repugnant to humans it would
likewise be repellent to annoying insects.

Prior to World War II, there were only four principal repellents: (1)
Oil of citronella, discovered in 1901, used also as a hair-dressing
fragrance by certain Eastern cultures; (2) dimethyl phthalate, discovered in 1929; (3) Indalone, introduced in 1937, and (4) Rutgers 612,
which became available in 1939.

With the onslaught of World War II and the introduction of American military personnel into new environments, particularly the
tropics, it became necessary to find new repellents that would survive both time and dilution by perspiration. The ideal repellent
would be nontoxic and nonirritating to humans, nonplasticizing,
and long-lasting (12 hours) against mosquitoes, biting flies, ticks,
fleas, and chiggers.

Unfortunately, the ideal repellent has still not been found. Some
repellents have unpleasant odors, require massive dosages, are oily
or effective only for a short time, irritate the skin, or soften paint and
plastics.

Insect repellents come in every conceivable formulation—undiluted, diluted in a cosmetic solvent with added fragrance, aerosols, creams, lotions, treated cloths to be rubbed on the skin, grease sticks, powders, suntan oils, and clothes-impregnating laundry emulsions. Regardless of the formulation, the periods of protection they offer vary with the chemical, the individual, the general environment, insect species, and avidity of the insect.

What happens to repellents once applied? Why aren't they effective longer than one or two hours? No single answer is satisfactory—but generally it is because they evaporate, are absorbed by the skin, are lost by abrasion of clothing or other surfaces, and are diluted by perspiration. Usually it's a combination of them all, and the only solution is a fresh application.

The following chemical structures are those of the most commonly found in today's repellents. Of these, diethyl toluamide (Delphene®, DEET) is by far superior to all others against biting flies and mosquitoes.

MGK® REPELLENT 11

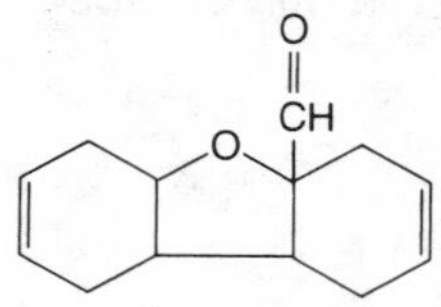

1,5α,6,9α9β-hexahydro-4α(4H)-dibenzofuran carboxaldehyde

MGK® REPELLENT 326

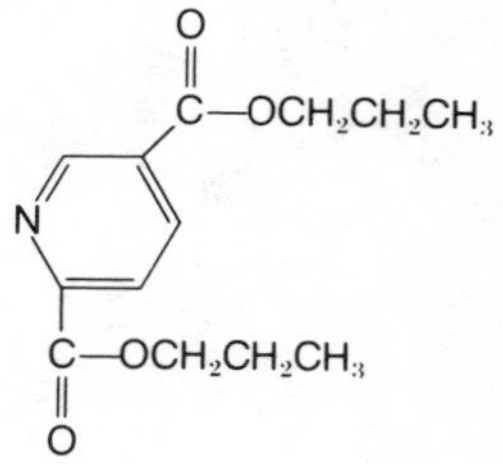

di-n-propyl 2,5-pyridinedicarboxylate

CLOTHING IMPREGNANT (ticks, chiggers)

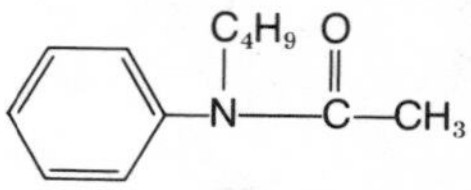

N-butyl acetanilide

MGK® REPELLENT 874 (ants, cockroaches)

$CH_3(CH_2)_7SCH_2CH_2OH$

2-(octylthio)ethanol

BENZYL BENZOATE (ticks, chiggers)

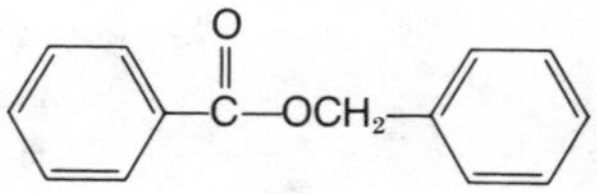

benzyl benzoate

STABILENE®, CRAG FLY REPELLENT (animals)

$C_4H_9O(C_3H_6O)_xH$

butoxy polypropylene glycol

TABATREX® (buildings only: ants, cockroaches, and flies)

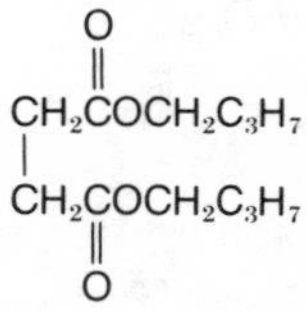

di-n-butylsuccinate

DIMETHYL PHTHALATE

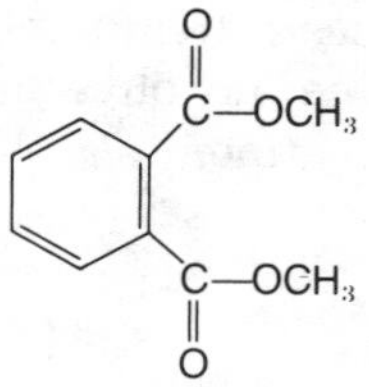

dimethyl 1,2-benzenedicarboxylate

DIBUTYL PHTHALATE

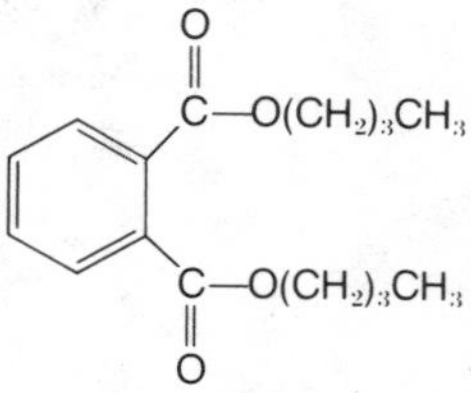

di-n-butyl phthalate

INDALONE

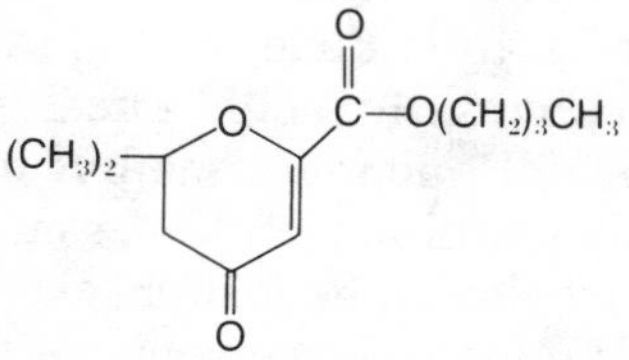

butyl 3,4-dihydro-2,2-dimethyl-4-oxo-2H-pyran-6-carboxylate

RUTGERS 612

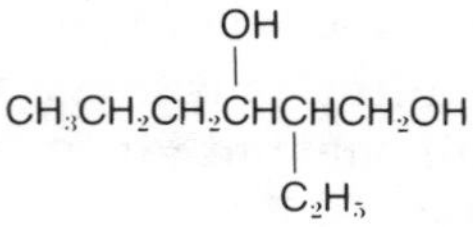

2-ethyl-1,3-hexanediol

STA-WAY®

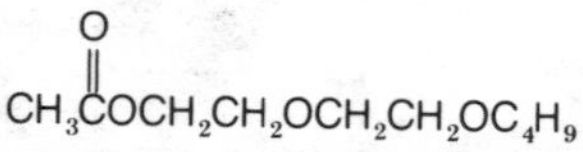

$CH_3COCH_2CH_2OCH_2CH_2OC_4H_9$

2-(2-butoxyethoxy)ethyl acetate

DIMETHYL CARBATE (Dimelone®)

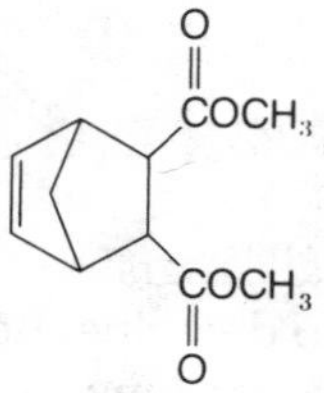

dimethyl cis-bicyclo(2,2,1)-5-heptane-2,3-dicarboxylate

DEET (Delphene®)

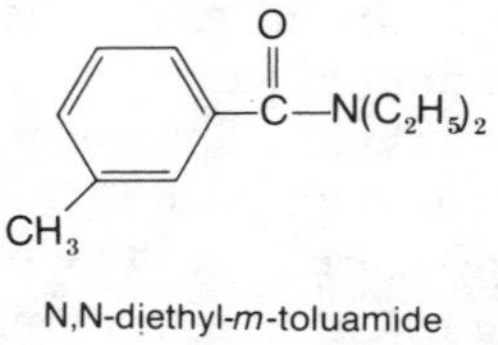

N,N-diethyl-m-toluamide

Animal repellents have been used for centuries, but with little success. One of the oldest known general repellents still in use is the plant resin asafetida, used as a medicinal and hung around the neck by the superstitious to ward off contagious diseases. In this respect, it may very well have been effective. It is so foul smelling that it undoubtedly serves also as a human repellent.

BIRD REPELLENTS

Bird repellents can be divided into three categories: (1) olfactory (odor), (2) tactile (touch), and (3) gustatory (taste). In the first category, only naphthalene granules or flakes are registered by the EPA. It should come as no surprise that naphthalene is repellent to all domestic animals as well.

Tactile repellents are made of various gooey combinations of castor oil, petrolatum, polybutane, resins, diphenylamine, pentachlorophenol, quinone, zinc oxide, and aromatic solvents applied as thin strips or beads to roosts, window ledges, and other favorite resting places.

The taste repellents are varied and somewhat surprising in certain instances, since they have other uses. The fungicides captan and copper oxalate are examples and are used as seed treatments to repel seed-pulling birds. Two other popular seed treatments are anthraquinone and glucochloralase. Turpentine, an old standby with multiple uses, can also be used as a seed treatment.

ANTHRAQUINONE

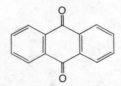

9,10-anthraquinone

CAPTAN

N-(trichloromethylthio)-4-cyclohexene-
1,2-dicarboximide

CHLORALOSE

glucochloralose

DOG AND/OR CAT REPELLENTS

The materials registered as dog and cat repellents are almost too numerous to describe. Both forms of moth crystals, naphthalene, and paradichlorobenzene, are readily available and very effective both indoors and outdoors.

Other materials found in commercial mixtures are: allyl isothiocyanate, amyl acetate, anethole, bittrex, bone oil, capsaicin, citral, citronella, citrus oil, creosote, cresylic acid, eucalyptus, geranium oil, lavender oil, lemongrass oil, menthol, methyl nonyl ketone, methyl salicylate, nicotine, pentanethiol, pyridine, sassafras oil, and thymol. The formulated product should be first checked out with your own nose if it's to be used indoors, since some of these substances are readily detectable and may be highly annoying. Better the dog or cat than such smells!

DEER REPELLENTS

Most of you urbanites are thinking how romantic it would be to live in a rural atmosphere and have deer nibbling around the cabin. To those who have the problem, it isn't all that great, particularly when the deer prune leaves and limbs from fruit and ornamental trees up to 8 feet above the ground by standing on their hind legs. Materials registered as effective deer repellents include two fungicides—thiram and ziram—and bone oil.

MOLE REPELLENTS

Only the fungicide thiram and liquid form of paradichlorobenzene are registered.

RABBIT REPELLENTS

These include blood dust, naphthalene, nicotine, and the two fungicides thiram and ziram.

RODENT REPELLENTS

The term rodent is rather all-inclusive and perhaps a bit deceptive, because not all rodent repellents are registered for all rodents. This requires the accurate identification of the particular pest and selecting one of the following materials or combinations that clearly indicates that pest on its label: Biomet-12 (tri-n-butyltin chloride), Endrin, naphthalene, paradichlorobenzene, polybutanes, polyethylene, R-55 (tert-butyl dimethyltrithioperoxycarbamate), and thiram.

NICOTINE

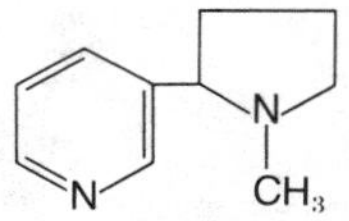

3-(1-methyl-2-pyrrolidyl) pyridine

THIRAM

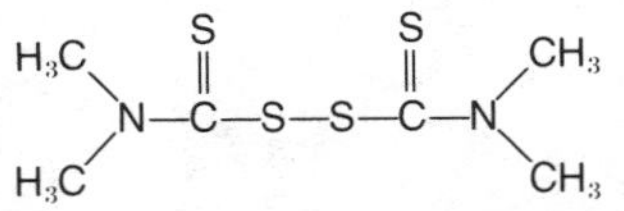

tetramethylthiuramdisulfide

ZIRAM

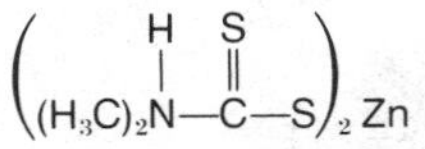

zinc dimethyldithiocarbamate

NAPHTHALENE

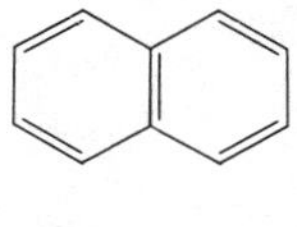

ENDRIN

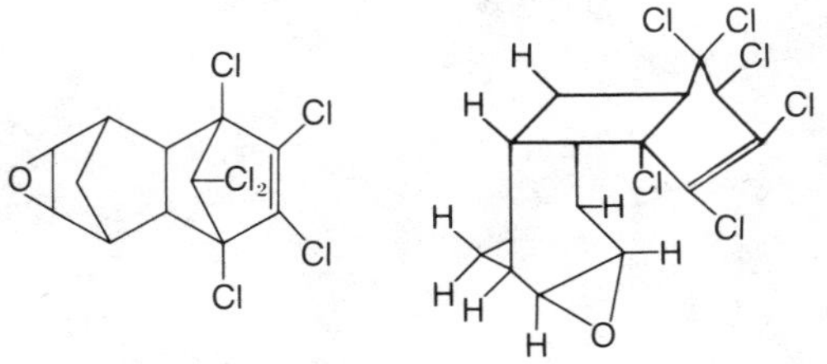

1,2,3,4,10,10-hexachloro-6,7-epoxy-1,4,4a,5,6,7,8,8a-octahydro-1,4-endo-endo-5,8-dimethanonaphthalene

PARA-DICHLOROBENZENE

BIOMET 12

$(C_4H_9)_3$—Sn—Cl

tri-n-butyltin chloride

R-55 REPELLENT®

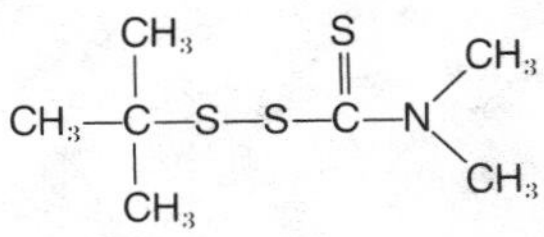

tertiary butylsulfenyl dimethyl-dithio carbamate

CHEMICALS USED IN
THE CONTROL OF PLANTS

CHAPTER 13

Herbicides

Now 'tis the spring, and weeds
are shallow rooted;
Suffer them now and they'll o'ergrow
the garden.

William Shakespeare, *King Henry VI*

Bowed by the weight of centuries he leans
Upon his hoe and gazes on the ground,
The emptiness of ages in his face,
And on his back the burden of the world.

Edwin Markham, *The Man with the Hoe*

The man with the hoe is no more. Herbicides, or chemical weed killers, have largely replaced mechanical methods of weed control in the past 30 years, especially where intensive and highly mechanized agriculture is practiced. Herbicides provide a more effective and economical means of weed control than cultivation, hoeing, and hand pulling. Together with fertilizers, other pesticides, and improved plant varieties, they have made an important contribution to the increased yields we now have and are one of the few ways remaining to combat rising costs and shortages of agricultural labor. The heavy use of herbicides is confined to North America, western Europe, Japan, and Australia. Without the use of herbicides, it would have been impossible to mechanize fully the production of cotton, sugar beets, grains, potatoes, and corn.

There are other locations where herbicides are used extensively. These include areas such as industrial sites, roadsides, ditch banks, irrigation canals, fence lines, recreational area, railroad embankments, and power lines. Herbicides remove undesirable plants that might cause damage, present fire hazards, or impede work crews. They also reduce costs of labor for mowing.

Herbicides are classed as *selective* when they are used to kill weeds without harming the crop and as *nonselective* when the purpose is to kill all vegetation. Both selective and nonselective materials can be applied to weed foliage or to soil containing weed seeds and seedlings, depending on the mode of action. The term *true selectivity* refers to the capacity of an herbicide, when applied at the proper dosage and time, to be active only against certain species of plants but not against others. But selectivity can also be achieved by placement, as when a nonselective herbicide is applied in such a way that it contacts the weeds but not the crop.

Herbicide classification would be a simple matter if only the selective and nonselective categories existed. However there are multiple-classification schemes that may be based on selectivity, contact versus translocation, timing, area covered, and chemical classification.

Each herbicide affects plants either by contact or translocation. Contact herbicides kill the plant parts to which the chemical is applied and are most effective against annuals, those weeds that germinate from seeds and grow to maturity each year. Complete coverage is essential in weed control with contact materials. And then there are the translocated herbicides, which are absorbed either by roots or above-ground parts of plants and then moved within the plant system to distant tissues. Translocated herbicides may be effective against all weed types; however, their greatest advantage is seen when used to control established perennials, those weeds that continue their growth from year to year. Uniform application is needed for the translocation materials, whereas complete coverage is not required.

Another method of classification is the timing of herbicide application with regard to the stage of crop or weed development. Applications may depend on many factors, including the chemical classification of the material and its persistence, the crop and its tolerance to the herbicide, weed species, cultural practices, climate, and soil type and condition. The three categories of timing are preplanting, preemergence, and postemergence.

Preplanting applications for control of annual weeds are made to an area before the crop is planted, within a few days or weeks of planting. Preemergence applications are completed prior to emergence of the crop or weeds, depending on definition, after planting. Postemergence applications are made after the crop or weed emerges from the soil.

Herbicide application based on area covered involves four categories: band, broadcast, spot treatments, and directed spraying. A band application treats a continuous strip, as along or in a crop row. Broadcast applications cover the entire area, including the crop. Spot treatments are confined to small areas of weeds. Directed sprays are applied to selected weeds or to the soil to avoid contact with the crop.

We come now to the chemical classification of herbicides, on which the major emphasis of this portion of the book is placed. The two major classifications are inorganic and organic herbicides.

THE INORGANIC HERBICIDES

The first chemicals utilized in weed control were inorganic compounds. There were brine and a mixture of salt and ashes, both of which were used to sterilize the land as early as Biblical times, by the Romans. In 1896, copper sulfate was used selectively to kill weeds in grain fields. From about 1906 until 1960, sodium arsenite solutions were the standard herbicides of commerce. Arsenic trioxide has been used at the incredible rates of 400 to 800 pounds per acre for soil sterilization. As with the arsenical insecticides, the trivalent arsenicals were nonspecific inhibitors of those enzymes containing sulfhydryl groups. They also uncouple oxidative phosphorylation.

Ammonium sulfamate ($NH_4SO_3NH_2$), another salt, was introduced in 1942 for brush control. Other salts have been used over the years. These include ammonium thiocyanate, ammonium nitrate, ammonium sulfate, iron sulfate, and copper sulfate, each applied heavily as a foliar spray. Their mechanisms of action are desiccation and plasmolysis (shrinkage of cell protoplasm away from its wall due to removal of water from its large central vacuole).

The borate herbicides were another family of inorganics, for example, sodium tetraborate ($Na_2B_4O_7 \cdot 5H_2O$), sodium metaborate ($Na_2B_2O_4 \cdot 4H_2O$), and amorphous sodium borate ($Na_2B_8O_{13} \cdot 4H_2O$). The amount of boron or boric acid determined their effectiveness. Borates are absorbed by plant roots, are translocated to above-ground parts, and are nonselective, persistent herbicides. Boron accumulates in the reproductive structures of plants, but its mechanism of toxicity is unclear. The borates are still used to give a semipermanent form of sterility to areas where no vegetation of any sort is wanted.

A nonselective herbicide used extensively for the last 40 years is sodiun   e ($NaClO_3$). It acts as a soil sterilant at rates of 200 to 1000 pounds per acre, but it can be used as a foliar spray at 5 pounds per acre as a defoliant of cotton. Caution must be taken with sprays of sodium chlorate to be certain the formulation contains flame retardants. Sulfuric acid has also been used as a foliar herbicide, but its use is limited considerably by its corrosiveness to metal spray rigs. Their mechanisms of action are those described for the miscellaneous salts, namely desiccation and plasmolysis.

Several of the inorganic herbicides are still useful in weed and brush control, but are gradually being replaced by organic materials. Although organic herbicides are not superior to inorganic ones, intensive EPA restrictions have been placed on some inorganic herbicides because of their persistence in soils. The inorganics are not wise choices for use around the home except by experts aiming to remove all vegetation from an area.

THE ORGANIC HERBICIDES

Petroleum Oils

The earliest organic herbicides were the petroleum oils, which are a complex mixture of long-chain hydrocarbons containing traces of nitrogen- and sulfur-linked compounds. These mixtures include alkanes, alkenes, and often alicyclics and aromatics. The petroleum oils are effective contact herbicides for all vegetation. Many homeowners today make use of old crankcase oil, gasoline, kerosene, and diesel oil for spot treatments. This is not a patriotic selection, in view of the drive to save energy and avoid adding hydrocarbons to our already overburdened atmosphere. Although only temporary in effect, they are fast-acting and very safe to use around the home. The petroleum oils exert their lethal effect by penetrating and disrupting plasma membranes.

Organic Arsenicals

Widely used as agricultural herbicides are the organic arsenicals, namely the arsinic and arsonic acid derivatives. Cacodylic acid (di-

ARSONIC ACID

$$O$$
$$R—As—OH$$
$$OH$$

ARSINIC ACID

$$O$$
$$R—As—OH$$
$$R'$$

MSMA

$$O$$
$$CH_3—As—ONa$$
$$OH$$

monosodium methanearsonate

CACODYLIC ACID

$$O$$
$$CH_3—As—OH$$
$$CH_3$$

hydroxydimethylarsine oxide

DSMA

$$O$$
$$CH_3—As—ONa$$
$$ONa$$

disodium methanearsonate

CACODYLIC ACID

$$O$$
$$CH_3—As—ONa$$
$$CH_3$$

sodium salt

2,4-D

$$Cl$$
$$Cl$$
$$OCH_2COH$$
$$O$$

(2,4-dichlorophenoxy)acetic acid

2,4,5-T

$$Cl$$
$$Cl$$
$$Cl$$
$$OCH_2COH$$
$$O$$

(2,4,5-trichlorophenoxy)acetic acid

ALLIDOCHLOR or CDAA

$$O$$
$$ClCH_2CN$$
$$CH_2CH=CH_2$$
$$CH_2CH=CH_2$$

N,N-diallyl-2-chloroacetamide

methylarsinic acid) and its sodium salt are the only derivatives of arsinic acid. Disodium methanearsonate (DSMA) and monosodium methanearsonate (MSMA) are salts of arsonic acid.

The organic arsenicals are much less toxic to mammals than the inorganic forms, are crystalline solids, and are relatively soluble in water. The arsonate, or pentavalent arsenic, acts in a way different from the trivalent form of the inorganic arsenicals just described. The arsonates upset plant metabolism and interfere with normal growth by entering into reactions in place of phosphate. Not only do they substitute for essential phosphate, but the arsonates are also absorbed and translocated in a manner similar to the way phosphates are absorbed and translocated.

The trivalent arsenates are exclusively contact herbicides, whereas the pentavalent arsonates are translocated to underground tubers and rhizomes, making them extremely useful against Johnson grass and nut sedges. They are usually applied as spot treatments.

Phenoxyaliphatic Acids

An organic herbicide was introduced in 1944, later to be known as 2,4-D, which was the first of the "phenoxy herbicides," "phenoxyacetic acid derivatives," or "hormone" weed killers. These were highly selective for broad-leaf weeds and were translocated throughout the plant. 2,4-D provided most of the impetus in the commercial search for other organic herbicides in the 1940s. There are several compounds belonging to this group, of which 2,4-D and 2,4,5-T are the most familiar. Other important compounds belonging to this group are 2,4-DB, MCPA, and silvex. Only 2,4-D and 2,4,5-T are illustrated.

The phenoxy herbicides have complex mechanisms of action resembling those of auxins (growth hormones) in some. They affect cellular division, activate phosphate metabolism, and modify nucleic acid metabolism.

2,4-D, MCPA, and 2,4,5-T have been used for years in gargantuan volume worldwide with no adverse effects on human or animal health. Recently, however, 2,4,5-T, used mainly for control of woody perennials, became the subject of heavy investigation, particularly because of its use in Vietnam. Certain samples were found to contain excessive amounts of a highly toxic impurity, tetrachlorodioxin, commonly referred to as *dioxin*. Slight alterations in manufacturing procedures brought this dioxin content down to acceptable levels after the source was determined.

Substituted Amides

The amide herbicides have diverse biological properties. These are simple molecules that are easily degraded by plants and soil. One of the earliest was allidochlor, which is selective for the grasses. Allidochlor inhibits the germination or early seedling growth of most annual grasses probably by alkylation of the —SH groups of proteins.

Diphenamid is used as a preemergence soil treatment and has little contact effect. Most established plants are tolerant to diphenamid, because it affects only seedlings. It persists from 3 to 12 months in soil.

Propanil has been used extensively on rice fields as a selective postemergence control for a broad spectrum of weeds.

There are many modes of actions for herbicides. One such, of great importance, is the inhibition of the Hill reaction. Propanil is a good example of such a herbicide, since it acts primarily in the leaves and is a strong inhibitor of the Hill reaction. This is a light-initiated reaction that splits water (photolysis), resulting in the production of free oxygen (O_2) by plants. Chlorophyll, the green pigment of plants, is an essential ingredient in the reaction, since it catalyzes the production of oxygen from water and the transfer of the hydrogen to a hydrogen acceptor. A simplified chemical formula for the reaction could be written as follows:

$$2H_2O + 2A \xrightarrow[\text{chlorophyll}]{\text{light}} 2AH_2 + O_2$$

where "A" is some unidentified hydrogen acceptor. The hydrogen and acceptor complex (AH_2) continue on in reactions with CO_2 to form plant sugars and cellulose, while free O_2 is released into the atmosphere.

We do not find a consistent pattern in the mechanism of action of amide herbicides. This is particularly noticeable when we recall that some are applied only to the soil and are active through the root system or seeds, while others are applied only to foliage, thus indicating different modes and sites of action.

Nitroanilines

Examples of the substituted dinitrotoluidines, which have high preemergence herbicidal action, are trifluralin and benefin. Trifluralin has very low water solubility, which minimizes leaching and movement away from the target. The nitroanilines inhibit both root and shoot growth when absorbed by roots, but they have an involved biochemical effect, which includes inhibiting the development of several enzymes and the uncoupling of oxidative phosphorylation.

Substituted Ureas

The substituted ureas, H_2NCNH_2, have their hydrogen atoms replaced with various carbon chain and ring structures to yield a utilitarian group of compounds used primarily as selective preemergence herbicides. The ureas are strongly adsorbed by the soil, then absorbed by roots. Their mechanism of action is inhibition of photosynthesis—production of plant sugars—and, indirectly, through inhibition of the Hill reaction.

DIPHENAMID

N,N-dimethyl-2,2-diphenylacetamide

PROPANIL

3',4'-dichloropropionanilide

TRIFLURALIN

α,α,α-trifluoro-2,6-dinitro-N,N-dipropyl-p-toluidine

BENEFIN

N-butyl-N-ethyl-α,α,α-trifluoro-2,6-dinitro-p-toluidine

MONURON

3-(p-chlorophenyl)-1,1-dimethylurea

DIURON

3-(3,4-dichlorophenyl)-1,1-dimethylurea

PROPHAM

isopropyl carbanilate

BARBAN

4-chloro-2-butynyl *m*-chlorocarbanilate

EPTC

C_2H_5S—C—$N(CH_2$—CH_2—$CH_3)_2$

S-ethyl dipropylthiocarbamate

PEBULATE

S-propyl butylethylthiocarbamate

TRIAZINE NUCLEUS

Carbamates

The esters of carbamic acid are physiologically quite active. As we have seen, some carbamates are insecticidal, and, as we shall see later, others are fungicidal. Still other carbamates are herbicidal, and we shall treat that group in this section. Discovered in 1945, the carbamates are used primarily as selective preemergence herbicides, but some are also effective as postemergence ones.

The first of the herbicidal carbamates was propham; it was followed first by chlorpropham, then by barban and terbutol. These carbamates kill plants by stopping cell division and plant tissue growth. Two effects are noted: cessation of protein production and shortening of chromosomes undergoing mitosis (duplication).

Propham and barban are aryl carbamates, because they all contain an aryl group, the six-member carbon ring, normally the phenyl ring; the alkyl group is the carbon chain or hydrocarbon radical with one hydrogen displacement by attachment of the alkyl group to the remainder of the molecule.

Another group within the carbamates are the thiocarbamates, which contains sulfur (from the Greek word *theion*) and which are

derived from the hypothetical thiocarbamic acid, HS—C—NH_2. The thiocarbamates are selective herbicides marketed for weed control in croplands. They are quite volatile and must be incorporated in the soil after application.

Heterocyclic Nitrogens

The triazines, which are six-member rings containing three nitrogens (the prefix *tri*- means "three") and azine (a nitrogen-containing ring) make up the heterocyclic nitrogens. The fundamental triazine nucleus is illustrated, showing the placement of the three nitrogens.

Probably the most familiar group of heterocyclic nitrogens, because of their heavy use and notoriety, are the triazines, which are strong inhibitors of photosynthesis. Their selectivity depends on the ability of tolerant plants to degrade or metabolize the parent compound (the susceptible plants do not). Triazines are applied to the soil primarily for their postemergence activity. There are many triazines on the market today, two of which are illustrated.

ATRAZINE

2-chloro-4-(ethylamino)-6-
(isopropylamino)-*s*-triazine

SIMAZINE

2-chloro-4,6-bis(ethylamino)-
s-triazine

AMITROLE

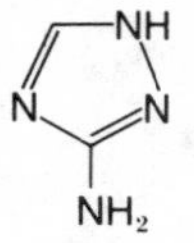

3-amino-s-triazole

PYRIDINE

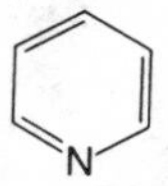

PICLORAM

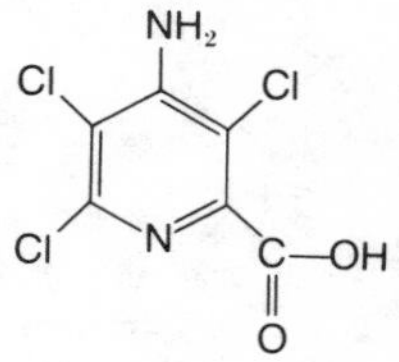

4-amino-3,5,6-trichloropicolinic acid

The triazoles are five-member rings containing three nitrogens. The outstanding member of this group is amitrole, the notorious aminotriazol that became embroiled in the historic "cranberry incident" of 1959. It was this event that culminated in the addition of the Delaney Amendment to the Pure Food and Drug Act in 1960 (See Chapter 22, "The Law and Pesticides.") In essence, this amendment stated that no residue of any carcinogen (cancer-producing agent) in a food crop would be tolerated. But more about that later. Amitrole acts in about the same way as the triazines, by inhibiting photosynthesis.

A popular herbicide derived from the pyridine molecule is picloram. Both are illustrated. Picloram is a readily translocated herbicide used against broad-leafed and woody plants; it may be taken up from either the roots or the foliage. It has also been used experimentally as a growth regulator at extremely low dosages on apricots, cherries, and figs. Its mechanism of action is probably the regulation of protein and enzyme synthesis in cells through its effects on nucleic acid synthesis and metabolism.

The substituted uracils give a wide range of grass and broad-leaf control for an extended period by inhibition of photosynthesis.

URACIL NUCLEUS

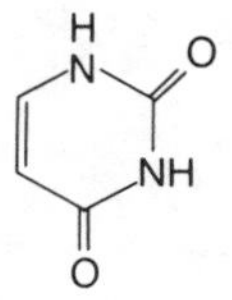

TERBACIL

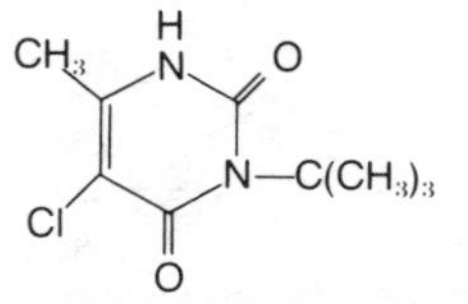

3-tert-butyl-5-chloro-6-methyluracil

Aliphatic Acids

Two heavily used herbicides in this group are TCA and dalapon, used against grasses, particularly our old enemies (or friends, depending on your view), quackgrass and Bermuda grass. They both act by precipitation of protein within the cells. Dalapon is widely used around homes to control Bermuda grass.

TCA

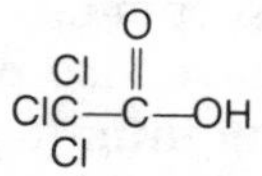

trichloroacetic acid

DALAPON

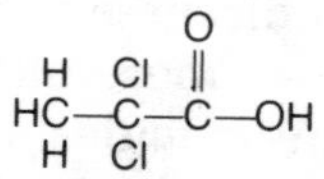

2,2,2-dichloropropionic acid

Arylaliphatic Acids

The arylaliphatic acids are aryls, or six-member rings, attached to aliphatic acids. A number of these materials are employed as herbicides and are applied to the soil against germinating seeds and seedlings. The mechanism of action for dicamba and fenac is not completely understood; however, it is probably similar to that of the phenoxy herbicides (2,4-D, etc.), which interfere with the nucleic acid metabolism.

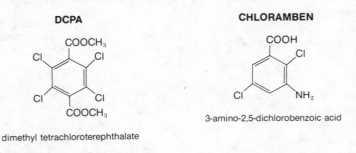

DICAMBA

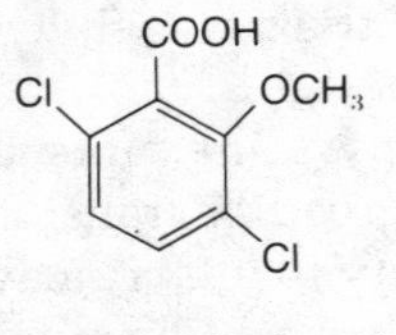

2-methoxy-3,6-dichlorobenzoic acid

FENAC

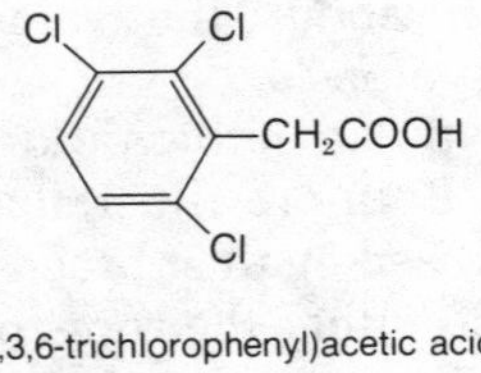

(2,3,6-trichlorophenyl)acetic acid

DCPA and chloramben produce auxinlike growth effects in plants but the available research data are insufficient to propose a mechanism of action.

DCPA

dimethyl tetrachloroterephthalate

CHLORAMBEN

3-amino-2,5-dichlorobenzoic acid

DNOC

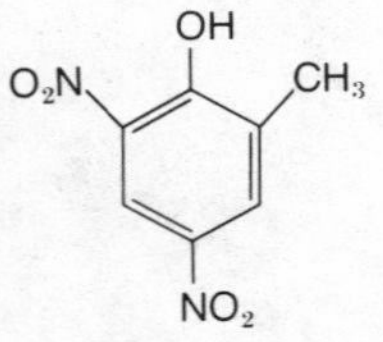

4,6-dinitro-o-cresol

DINOSEB

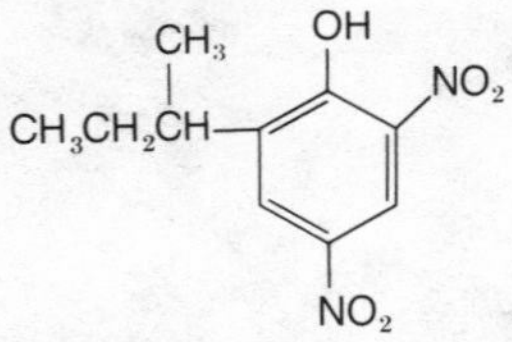

2-sec-butyl-4,6-dinitrophenol

PCP

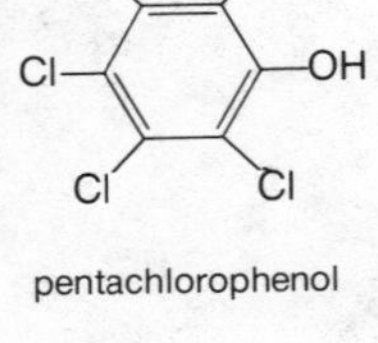

pentachlorophenol

Phenol Derivatives

Phenol derivatives are highly toxic to humans by every route of entry into the body and are nonselective foliar herbicides that are most effective in hot weather. The nitrophenols, represented by DNOC, were first introduced as herbicides in 1932. Dinoseb is the best-known representative of the nitrophenols, which act by uncoupling oxidative phosphorylation.

The dinitrophenols are familiar compounds, since they have already been discussed in the section on the insecticides. As you will recall, one or more members of this group have also been used as ovicides, insecticides, fungicides, and blossom-thinning agents.

Another subgroup of the phenol derivatives is the chlorinated phenols, of which only one member is recognized as a herbicide, namely PCP or pentachlorophenol, which is also used for termite protection, wood treatment for fungal rots, nonselective herbicide, and as a preharvest defoliant. Its mechanism is a combination of plasmolysis, protein precipitation, and desiccation. Because of its wide effectiveness and multiple routes of action, PCP is recognized as being destructive to all living cells.

Substituted Nitriles

Nitriles are organic compounds containing the C≡N or cyanide grouping. There are several substituted nitrile herbicides, which have a wide spectrum of uses against grasses and broad-leaf weeds. Their mechanisms of action are broad, involving seedling growth inhibition, potato sprout inhibition, and gross disruption of tissues by inhibiting oxidative phosphorylation and preventing the fixation of CO_2. These effects, of course, do not explain their rapid action. This is attributed primarily to rapid permeation and release of C≡N⁻.

DICHLOBENIL

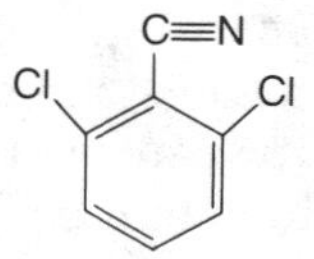

2,6-dichlorobenzonitrile

BROMOXYNIL

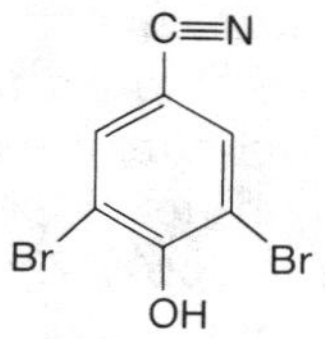

3,5-dibromo-4-hydroxybenzonitrile

Bipyridyliums

The name *bipyridylium* suggests the attachment of two pyridyl rings. There are two important herbicides in this group, diquat and paraquat. Both are contact herbicides that damage plant tissues quickly, causing the plants to appear frostbitten because of cell membrane destruction. Both materials reduce photosynthesis and are more effective in the light than in the dark. Neither material is active in soils. They are available only to professional weed control specialists, who achieve spectacular results for homeowners.

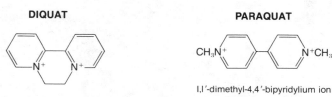

DIQUAT

6,7-dihydrodipyrido(1,2-α:2′,1′-c)pyrazidinium

PARAQUAT

l,l′-dimethyl-4,4′-bipyridylium ion

Miscellaneous Herbicides

Within the miscellaneous herbicides belongs methyl bromide (CH_3Br), which is used as a fumigant for any known organism in soil—nematodes, fungi, seed, insects, and other plant parts. Allyl alcohol (CH_2=$CHCH_2OH$) is a volatile, water-soluble fumigant used for the same purposes as methyl bromide.

Endothall is used as an aquatic weed and a selective field crop herbicide. It acts by interfering in RNA synthesis. Endothall has one

ENDOTHALL

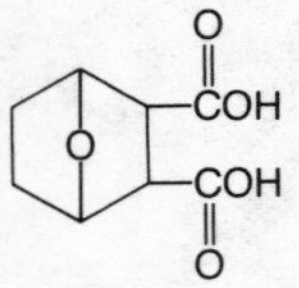

7-oxabicyclo (2,2,1)heptane-2,3-di-
carboxylic acid, sodium salt

BENSULIDE

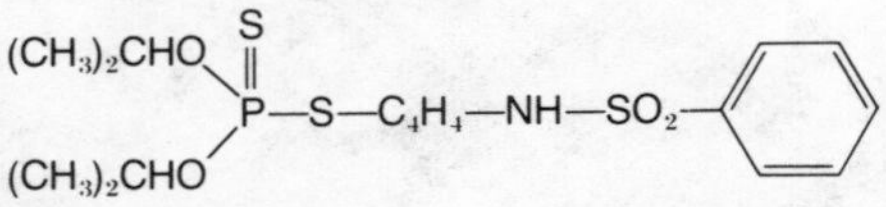

O,O-diisopropyl phosphorodithioate S-ester with
N-(2-mercaptoethyl)benzenesulfonamide

ACROLEIN

$$CH_2=CH-CHO$$

distinct advantage over most aquatic weed killers—its low toxicity to fish, an outstanding example of environmental protection through pesticide selectivity.

One of the better turf herbicides, especially for the control of crabgrass, is bensulide. It is an organophosphate, but is considered one of the less toxic herbicidal materials. Bensulide acts by inhibiting cell division in root tips. It is used as a preemergence herbicide in lawns to control certain grasses and broad-leaf weeds, but it fails as a foliar spray because it is not translocated.

Acrolein requires application only by licensed operators because of its frightening tear-gas effect, but it is an extremely useful aquatic weed herbicide. Plants exposed to acrolein disintegrate within a few hours and float downstream. It is a general plant toxicant, destroying plant cell membranes and reacting with various enzyme systems. By using spot treatment in lakes, the fish population can be saved. This is another example of environmental protection, but by a different method, spot application.

Many chemical and use classes of herbicides are available, some of which have different mechanisms of action within the same chemical class. Those discussed and illustrated here are but a cross-section of the existing herbicides. We can expect to see different classes develop in the future, just as in the past, with mechanisms of action more clearly delineated, following intensive research on this complex subject. Because of the wealth of materials and action, no good summary can be made. Rather, the reader is urged to review the groups frequently, to establish generalizations regarding herbicide use and action classes.

CHAPTER 14

Plant Growth Regulators

They came ... and cut down from there a branch with a single cluster of grapes, and they carried it on a pole between two of them.

Numbers 13:23

Chemicals used in some manner to alter the growth of plants, blossoms, or fruits are plant growth regulators (PGRs) and are legally considered pesticides. Plant growth regulators are also referred to as *plant regulating substances, growth regulants, plant hormones,* and *plant regulators.*

Natural substances are produced by plants that control growth, initiate flowering, cause blossoms to fall, set fruit, cause fruit and leaves to fall, control initiation and termination of dormancy, and stimulate root development. These natural substances are hormones.

The history of PGRs begins in 1932, when it was discovered that acetylene and ethylene would promote flowering in pineapples. In 1934, auxins were found to enhance root formation in cuttings. Since then, outstanding developments have occurred as a consequence of the use of PGRs. Some of the major discoveries were the development of seedless fruit; prevention of berry and leaf drop in holly; prevention of early, premature drop of fruit; promotion of heavy setting of fruit blossoms; thinning of blossoms and fruit; prevention of sprouting in stored potatoes and onions; and the inhibition of buds in nursery stock and fruit trees to prolong dormancy. These are just a few of the hundreds of achievements made possible with these growth-controlling chemicals. As a result, we eat better-quality foods, eat fruits and vegetables "out-of-season" or literally year-round, and pay less for our food because of the reduced need for hand labor in thinning and harvesting.

Six classes of PGRs are recognized by the American Society for Horticultural Science: auxins, gibberellins, cytokinins, ethylene generators, inhibitors, and growth retardants. These very useful agricultural chemical tools will be discussed under these headings.

AUXINS

Auxins are compounds that induce elongation in shoot cells. Some occur naturally, whereas others are manufactured. Auxin precursors are materials that are metabolized to auxins in the plant. Antiauxins are chemical compounds that inhibit the action of auxins.

Auxins are used to thin apples and pears, increase yields of potatoes, soybeans, and sugar beets, to assist in the rooting of cuttings, and to increase flower formation, among other things. The mechanism of action is not completely understood, but they may work by controlling the type of enzyme produced in the cell. In any event, with the addition of auxin the individual cells become larger by a loosening of the cell wall, which is followed by the increased uptake of water and expansion of the cell wall.

IAA

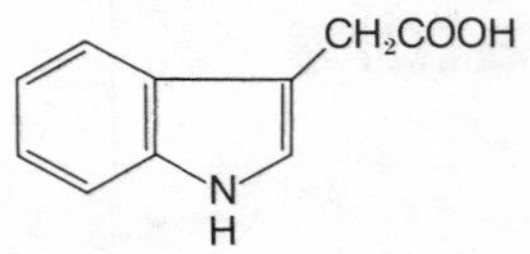
CH₂COOH
N
H

indoleacetic acid

2,4-D

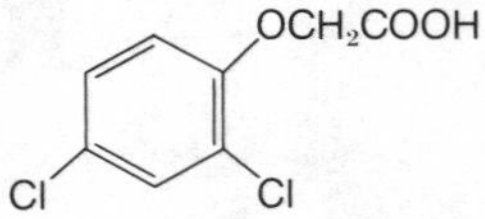
OCH₂COOH
Cl
Cl

2,4-dichlorophenoxyacetic acid

MCPB

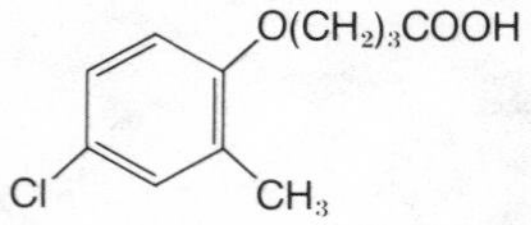
O(CH₂)₃COOH
Cl
CH₃

4-[(4-chloro-*o*-tolyl)oxy]butyric acid

IAAld

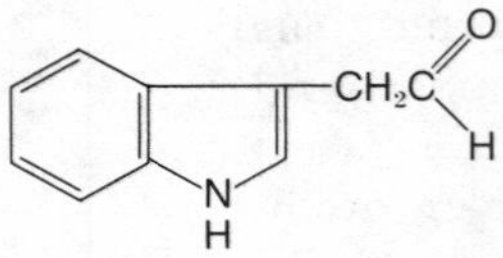
O
CH₂C
H
N
H

indoleacetaldehyde

BNAA

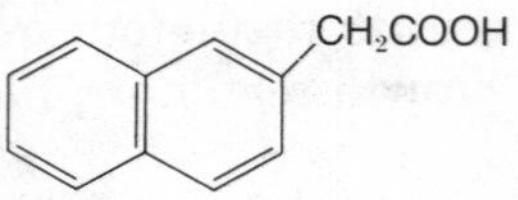
CH₂COOH

β-naphthaleneacetic acid

BNOA

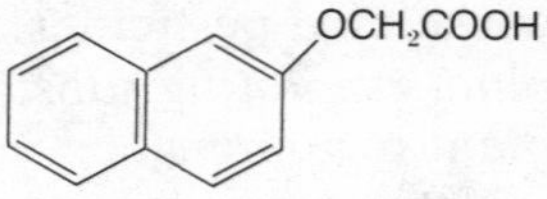
OCH₂COOH

β-naphthoxyacetic acid

GIBBERELLINS

Gibberellins are compounds that stimulate cell division or cell elongation or both, and have a gibbane skeleton. In 1957, gibberellin, or gibberellic acid, was introduced to the horticultural world. It caused incredible growth in many types of plants. Although originally isolated from a fungus, gibberellins were later found to be natural constituents in all plants. Since then, more than 33 gibberellins have been isolated and are identified as GA_1, GA_2, etc. The gibberellic acid most commonly used is GA_3. The mechanism of action for gibberellins is the induction or manufacture of more enzyme(s) in the cells, resulting in cell growth, particularly by elongation. The most striking effect of treating with gibberellin is the stimulation of growth, expressed as long stems.

Examples of gibberellins' uses are: to increase stalk length and yields of celery, to break dormancy of seed potatoes, to increase grape size, to induce seedlessness in grapes, to improve the size of greenhouse-grown flowers, to delay fruit maturity, to extend fruit harvest, and to improve fruit quality.

GIBBANE

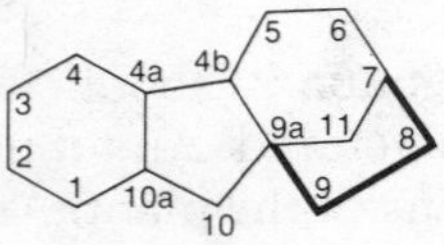
1
2
3
4
4a
4b
5
6
7
8
9
9a
10
10a
11

GA_3

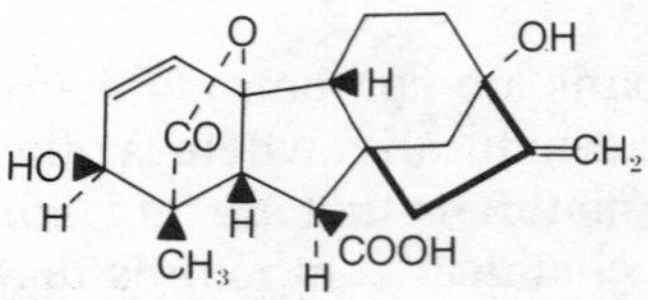
O
OH
H
CO
HO
CH₂
H
H
CH₃
H
COOH

gibberellic acid

CYTOKININS

Cytokinins (also referred to as *phytokinins*) are naturally occurring or manufactured compounds that induce cell division in plants. Most of the cytokinins are derivatives of adenine. These useful materials were discovered in 1955. Their practical potential lies in prolonging the storage life of green vegetables, cut flowers, and mushrooms.

Cytokinins cause two outstanding effects in plants—the induction of cell division and the regulation of differentiation in removed plant parts. Their mechanism of action is not known, but they apparently act at the gene level, becoming incorporated in the nucleic acids of the cell. This ultimately influences cell division, in which the nucleic acids play critical roles.

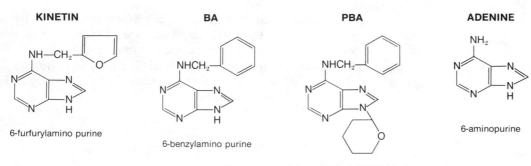

Naturally Occurring Cytokinins

ZEATIN

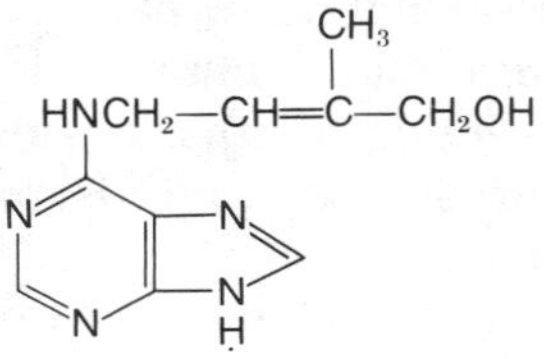

2iP

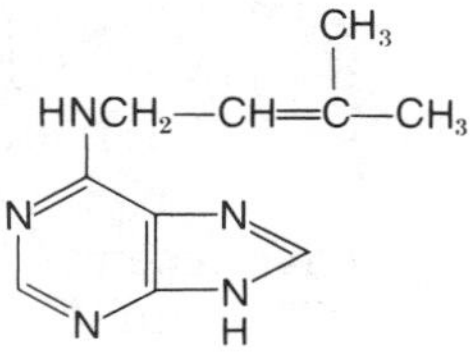

6-(γ,γ-dimethylallylamino)purine

Synthetic Cytokinins

KINETIN

6-furfurylamino purine

BA

6-benzylamino purine

PBA

6-(benzylamino)-9-(2-tetrahydro-pyranyl)-9*H*-purine

ADENINE

6-aminopurine

ETHYLENE GENERATORS

Do you recall that the first mention of a plant growth regulator was the use of ethylene in promotion of pineapple flowering? Recently, materials have been developed that are applied as sprays to the growth sites of plants and that stimulate the release of ethylene ($H_2C=CH_2$). Ethylene produces numerous physiological effects and can be used to regulate different phases of plant metabolism, growth, and development.

The ethylene generators also have many uses. They can be used to accelerate pineapple maturity; to induce uniform fig ripening; to facilitate mechanical harvest of peppers, cherries, plums, and

ETHEPHON

$$ClCH_2CH_2\overset{\displaystyle O}{\underset{\displaystyle OH}{P}}OH$$

2-(chloroethyl)phosphonic acid

Naturally Occurring Inhibitors

BENZOIC ACID

COOH

GALLIC ACID

HO — COOH, HO, HO, OH

CINNAMIC ACID

CH=CHCOOH

(S)-ABSCISIC ACID

2-TRANS-ABSCISIC ACID

apples; and to induce uniform flowering and fruit thinning, among other possibilities.

Ethylene can be used to stimulate seed germination and sprouting; abscission of flowers, leaves, and fruit; regulation of growth; and ripening of fruit, when introduced at the proper time. None of these biological effects results from a clearly defined mechanism of action. Generally, however, ethylene apparently has its greatest influence on the dominant enzymes of the particular physiological state of the absorbing tissue. The ethylene apparently serves as a trigger or synergist, resulting in a chain of premature biochemical events, expressed in the final and usually desirable result. Ethephon gradually releases ethylene as a degradation product when applied to plant surfaces.

INHIBITORS AND RETARDANTS

These are an assorted group of substances that inhibit a physiological process in plants. Those that occur naturally in plants are usually hormones and inhibit different functions; for instance, growth or seed germination, or the action of other hormones, gibberellins, and auxins. New types of synthetic compounds, plant growth retardants, have recently been discovered.

The inhibitors and retardants have many uses. They prevent sprouting of stored onions, potatoes, and other root crops; they retard sucker development on tobacco plants; they induce shortening of stems; they favor redistribution of dry matter; they prevent lodging of grain; and they permit controlled growth of flower crops and ornamentals.

These materials are a diverse group of compounds and thus have different biological effects on plants. They are antagonistic to the growth-promoting hormones, such as auxins, gibberellins and cytokinins, through a potential multitude of biochemical actions. We leave the plant growth regulators with this veil of mystery surrounding their curious and beneficial effects.

Synthetic Inhibitors

CCC—CYCOCEL

$$Cl—CH_2—CH_2—\overset{\displaystyle CH_3}{\underset{\displaystyle CH_3}{N^+}}—CH_3Cl^-$$

(2-chloroethyl)trimethyl ammonium chloride

MH—MALEIC HYDRAZIDE

1,2-dihydro-3,6-pyridazinedione

CHAPTER 15

Defoliants and Desiccants

Certain trees are more susceptible to ... arsenic than others. After each rain the poison ... is absorbed more, or is more active when wet, and ... it acts by dehydrating thereafter.

C. V. Riley, *U.S. Agricultural Report* (1886)

Chemical defoliation is the premature abscission of leaves, brought on by the formation of the abscission layer at the point where the leaf petiole joins the plant stem. Defoliants facilitate harvest operations by accelerating leaf fall from crop plants such as cotton, soybeans, or tomatoes. The defoliation of cotton is a good example. The premature removal of leaves from the cotton plant permits earlier harvesting, and results in higher grades of cotton because few leaves remain to clog the mechanical picker, add trash, or stain the fibers. Defoliation often helps lodged (fallen-over) plants to straighten up, increasing the plant exposure to sun and air. This enables the plants to dry quickly and thoroughly, and the mature bolls open faster, reducing boll rots that damage fiber and seed. And, finally, defoliation of the cotton plant reduces populations of fiber-staining insects, particularly aphids and whiteflies, which deposit honeydew in the open bolls.

Two conditions must exist before defoliation of any crop is effective: First the plant must be in a state of maturity in which growth has stopped, and second, the temperatures must exceed 80° during the day and 50° at night.

Chemicals that are used to speed the drying of crop plant parts such as cotton leaves and potato vines are called *desiccants*. They usually kill the leaves rapidly, freezing them to the plant and thus mimicking the effect of a light frost. Desiccants cause foliage to lose water. Leaves, stems, and even branches of plants are sometimes killed so rapidly by desiccants that an abscission layer has insufficient time to develop, and the drying leaves thus remain attached to the plant. The advantage of desiccants over defoliants is that they can be applied at a later date than defoliants. Thus additional time is gained, during which the leaves continue to function and contribute to seed and fiber quality, at least with cotton.

DEFOLIANTS

Inorganic Salts

The inorganic salts are the oldest defoliants. Sodium chlorate ($NaClO_3$), magnesium chlorate ($Mg(ClO_3)2 \cdot 6H_2O$), disodium octaborate tetrahydrate ($Na_2B_8O_{14} \cdot 4H_2O$), and the other sodium polyborates are good examples of these old standbys, which are still used, primarily on cotton. These are contact materials that, by virtue of their high acidity, bring about rapid destruction of the delicate protoplasmic structures, resulting in the formation of the petiole abscission layer. The exact nature and sequence of the chemical reactions is unknown. The chlorates cause chlorosis of leaves and a starch depletion in stems and roots when applied in less than lethal doses.

Aliphatic Acids

As indicated in the chapter on herbicides, an aliphatic acid is a carbon chain acid. Their sodium salts are also included in this group. Neodecanoic acid is a readily degradable compound of moderate effectiveness for use on most crops requiring defoliation.

Prep® is the sodium salt of an aliphatic acid, is used as a defoliant for cotton and potatoes, and is rapidly absorbed but not translocated.

Paraquat

Paraquat—also studied in the herbicides, under the dipyridylium class—damages plant tissue very rapidly. Its swift action results from the breakdown of plant cells responsible for photosynthesis, giving the leaves a waterlogged appearance within a few hours of treatment. Paraquat results in the formation of OH^- radicals or hydrogen peroxide (H_2O_2) as the primary toxicant(s). Because most leaves drop off, paraquat is considered a defoliant.

Organophosphates

Folex and DEF are two organophosphate defoliants that have proved extremely useful in cotton production. Neither of these compounds is hormone acting. Rather, they induce abscission by injuring the leaf, causing, in the levels of naturally occurring plant hormones, changes that induce the early formation of the leaf abscission layer. Defoliation follows treatment by 4 to 7 days.

As you have seen, only a small number of compounds are registered as defoliants. There are reasons for this. First, only a few crops require defoliation. Second, the materials must act rapidly to bring about abscission within a minimum time after application. And, third, the compounds must break down rapidly, leaving no undesirable residue in the target portion of the crop. Defoliants are not for lawn and garden use.

NEODECANOIC ACID

$(CH_3)_3C(CH_2)_5COH$

PREP®

$Cl(CH_2)_2C-ONa$

sodium *cis*-3 chloroacrylate

PARAQUAT

CH_3N^+ ... N^+CH_3

1,1'-dimethyl-4,4'-bipyridylium ion

TRIBUTYL PHOSPHOROTHIOITE (Folex®)

$(C_4H_9S)_3P$

S,S,S-tributyl phosphorotrithioite

TRIBUTYL PHOSPHOROTHIOATE (DEF®)

$(C_4H_9S)_3P{=}{=}O$

S,S,S-tributyl phosphorotrithioate

DESICCANTS

The Inorganics

The number of desiccants available is much larger than the number of defoliants. Ammonia and ammonium nitrate are both used as desiccants for cotton (they also ultimately add nutrients to the soil after the crop residue is returned to the soil). Petroleum solvents are applied to alfalfa and clover seed crops as well as to potatoes to enhance leaf drying and harvest.

Two of the inorganic salts mentioned in earlier chapters are used as desiccants for cotton—sodium borate(s) and sodium chlorate.

Arsenic acid is probably the oldest of the cotton desiccants, and it is still used in quantity. The material penetrates the leaf cuticle and the rapid contact injury precludes any extensive translocation. Arsenic acid uncouples oxidative phosphorylation and forms complexes with sulfhydryl-containing enzymes.

ARSENIC ACID

H_3AsO_4

Phenol Derivatives

Dinoseb and pentachlorophenol, both phenol derivative herbicides of universal effectiveness, are also used as desiccants for cotton. Pentachlorophenol has additional uses on seed crops of alfalfa, clovers, lespedeza, and vetch.

DINOSEB

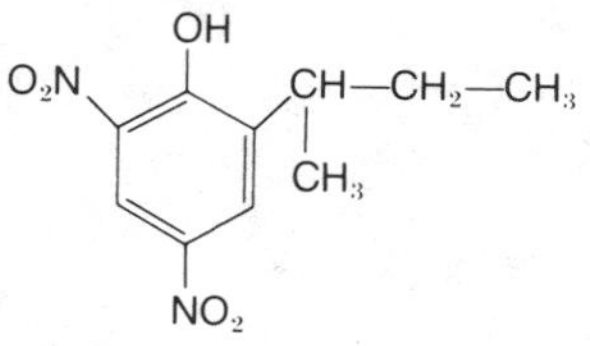

2-sec-butyl-4,6-dinitrophenol

PCP

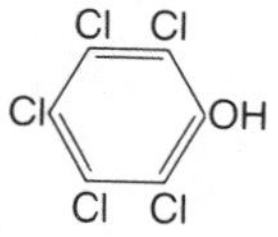

pentachlorophenol

Bipyridyliums

The bipyridylium herbicides, diquat dibromide and paraquat, because of their frostlike effect on green foliage, are outstanding desiccants. Diquat dibromide is used on the seed crops only of alfalfa, clover, soybeans, and vetch, while paraquat is used on cotton, potatoes, and soybeans.

DIQUAT

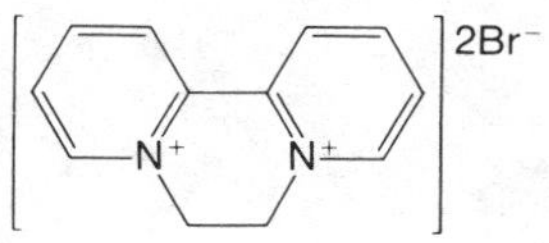

6,7-dihydrodipyrido(1,2-α : 2′,1′-c)
pyrazinediium ion

PARAQUAT

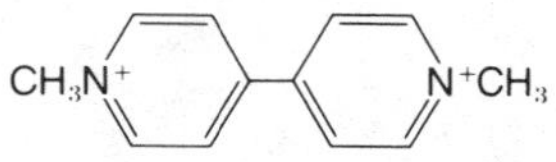

1,1′-dimethyl-4,4′-bipyridylium ion

Miscellaneous Desiccants

Endothall, examined briefly toward the end of the chapter on herbicides, is also a desiccant. It is used on cotton and—for the seed crops only—alfalfa, clovers, soybeans, trefoil, and vetch. It kills the leaves by contact through rapid penetration of the cuticle and results in desiccation and browning of the foliage. Its mode of action is not understood.

ENDOTHALL

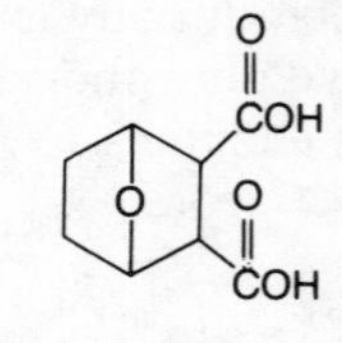

7-oxabicyclo (2,2,1)heptane-2,3-dicarboxylic acid

Ametryn, one of the very old and popular triazine herbicides, is also used as a potato vine desiccant prior to digging the potatoes. Ametryn penetrates the leaves rapidly and is a strong inhibitor of photosynthesis, causing the leaves to desiccate within 72 hours after application.

AMETRYN

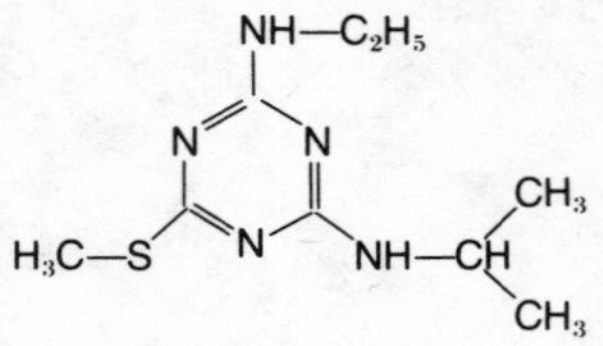

2-(ethylamino)-4-(isopropylamino)-
6-(methylthio)-S-triazine

PART V

CHEMICALS USED
IN THE CONTROL
OF MICROORGANISMS

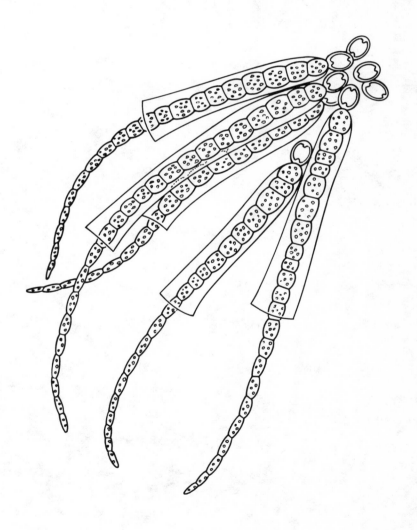

CHAPTER **16**

Fungicides and Bactericides

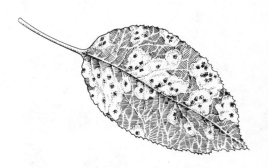

The Lord shall smite thee ...
with blasting and with mildew.
Deuteronomy 28:22.

I have smitten you with blasting
and with mildew; I laid waste
your gardens and your vineyards;
and your fig and your olive trees.
Amos 4:9.

Fungicides, strictly speaking, are chemicals used to kill or halt the development of fungi. However, for our purposes we shall consider them as chemicals used to control bacterial as well as fungal plant pathogens, the causal agents of most plant diseases. Other organisms that cause plant disease are viruses, rickettsias, algae, nematodes, mycoplasmalike organisms, and parasitic seed plants. We will deal only with chemicals used for the control of fungi and bacteria.

There are hundreds of examples of plant diseases. These include storage rots, seedling diseases, root rots, gall disease, vascular wilts, leaf blights, rusts, smuts, mildews, and viral diseases. These can, in many instances, be controlled by the early and continued application of selected fungicides that either kill the pathogens or inhibit their development.

Most plant diseases can be controlled to some extent with today's fungicides. Among those that cannot be controlled with chemicals are *Phytophthora* and *Rhizoctonia* root rots, *Fusarium*, *Verticillium*, and bacterial wilts, and the viruses. The difficulties with these diseases are that they occur either below ground and thus beyond the reach of fungicides, or they are systemic within the plant.

The fungal diseases are basically more difficult to control with chemicals than are insects, because the fungus is a plant living in close quarters with its host. This explains the difficulty of finding chemicals that kill the fungus without harming the plant. Also, fungi that can be controlled by fungicides may undergo secondary cycles rapidly and produce from 12 to 25 "generations" during a 3-month growing season. Consequently, repeated applications of protective fungicides may be necessary, due to plant growth dilution and removal by rain and other weathering.

It is necessary that fungicides be applied to protect plants during stages when they are vulnerable to inoculation by pathogens, before there is any evidence of disease. Fungicides, referred to as *chemotherapeutants*, can help to control certain diseases after the symptoms appear. Also, protective fungicides are commonly used, even after symptoms of disease have appeared. Eradicant fungicides are usually applied directly to the pathogen during its "overwintering" stage, long before disease has begun and symptoms have appeared. In the case of crops whose sale depends on appearance, such as lettuce and celery, however, the fungicide must be applied as a protectant spray, in advance of the pathogens, to prevent the disease.

There are about 150 fungicidal materials in our present arsenal, most of which are recently discovered organic compounds. Most of these act as protectants, preventing spore germination and subsequent fungal penetration of plant tissues. Protectants are applied repeatedly to cover new plant growth and to replenish the fungicide that has deteriorated or has been washed off by rain.

The application principle for fungicides differs from that of herbicides and insecticides. Only that portion of the plant that has a coating of dust or spray film of fungicide is protected from disease. Thus a good uniform coverage is essential. Fungicides, with several new exceptions, are not systemic in their action. They are applied as sprays or dusts, but sprays are preferable, since the films stick more readily, remain longer, can be applied during any time of the day, and result in less off-target drift.

Thanks to modern chemistry, many of the serious diseases of grain crops are controlled by treating the seeds with selective materials. Others are controlled with resistant varieties. Diseases of fruit and vegetables are often controlled by sprays or dusts of fungicides.

Historically, fungicides have centered around sulfur, copper, and mercury compounds, and even today most of our plant diseases could be controlled by these groups. However, the sulfur and copper compounds can retard growth in sensitive plants, and therefore the organic fungicides were developed. These sometimes have greater fungicidal activity and usually have less phytotoxicity.

The general-purpose fungicides for agriculture include inorganic forms of copper, sulfur (and mercury, until recently), and metallic complexes of cadmium, chromium, and zinc, along with a variety of organic compounds. The general-purpose lawn and garden fungicides are few in number and are usually organic compounds.

This brings us to the fungicides themselves, which can best be explored historically. We begin with inorganic fungicides.

THE INORGANIC FUNGICIDES

Sulfur

The control of plant diseases undoubtedly precedes written history and includes incantations to the gods. Virgil in his "Georgics" refers to seed treatment: "I have seen some medicate their seed before they sow it. They steep it in nitre and amurca to obtain a fuller produce in the deceitful pods" (Martyn, 1811, p. 52–53, Book I, 190–200).

Sulfur in many forms is probably the oldest effective fungicide known and is still a very useful garden fungicide. There are three physical forms or formulations of sulfur used as fungicides. The first is finely ground sulfur dust that contains 1 to 5 percent clay or talc to assist in dusting qualities. The sulfur in this form may be used as a carrier for another fungicide or an insecticide. The second is flotation or colloidal sulfur, which is so very fine that it must be formulated as a wet paste in order to be mixed with water. In its original, dry, microparticle size, it would be impossible to mix with water, but would merely float. Wettable sulfur is the third form; it is finely ground with a wetting agent so that it will mix readily with water for spraying. The easiest to use, of course, is dusting sulfur, applied when plants are slightly moist with the morning dew.

Certain pathogenic organisms and most mites are killed by direct contact with sulfur and also by its fumigant action at temperatures above 70°F. The fumigant effect is, however, somewhat secondary at marginal temperatures and under windy conditions. It is quite effective in controlling powdery mildews of plants that are not unduly sensitive to sulfur. Unlike those of any other fungus, spores of powdery mildews will germinate in the absence of a film of water in the penetration court (a spot on the tissue that is "softened" prior to penetration by spore). Its fumigant effect—acting at a distance—is undoubtedly important in killing spores of powdery mildews. Sulfur interferes in electron transport along the cytochromes and is then reduced to hydrogen sulfide (H_2S), a toxic entity to most cellular proteins.

Copper

The majority of inorganic copper compounds are practically insoluble in water and are pretty blue, green, red, or yellow powders sold as fungicides. The various forms include Bordeaux mixture, named after the Bordeaux region in France, where it originated. Bordeaux is a chemically undefined mixture of copper sulfate and hydrated lime, which was accidentally discovered when sprayed on grapes in Bordeaux to scare off "freeloaders." It was soon observed that downy mildew, a disease of grapes, disappeared from the treated plants. From this unique origin began the commercialization of fungicides. The copper ion, which becomes available from both the highly soluble and relatively insoluble copper salts, provides the fungicidal as well as phytotoxic and poisonous properties. A few of the many inorganic copper compounds used over the years are presented in Table 4.

A comment on solubility is appropriate at this point. In general, protective fungicides have low ionization constants, but in water some toxicant does go into solution. That small quantity absorbed by the fungal spore is then replaced in solution from the residue. The spore accumulates the toxic ion and "commits suicide," so to speak. Except for powdery mildews, water in the penetration court thus permits spore germination and makes soluble the toxic portion of the fungicidal residue.

The copper ion is toxic to all plant cells and must be used in discrete dosages or in relatively insoluble forms, to prevent killing all or portions of the host plant. This is the basis for the use of

TABLE 4
A sampling of the vast number of inorganic copper compounds used as fungicides.

Name	Chemical formula	Uses
Cupric sulfate	$CuSO_4 \cdot 5H_2O$	Seed treatment and preparation of Bordeaux mixture
Copper dihydrazine sulfate	$CuSO_4(N_2H_5)_2SO_4$	Powdery mildew Black spot of roses
Copper oxychloride	$3Cu(OH)_2 \cdot CuCl_2$	Powdery mildews
Copper oxychloride sulfate	$3Cu(OH)_2 \cdot CuCl_2$ and $3Cu(OH)_2 \cdot CuSO_4$	Many fungal diseases
Copper zinc chromates	$15CuO\text{-}10ZnO \cdot 6CrO_3 \cdot 25H_2O$	Diseases of potato, tomato, cucurbits, peanuts, and citrus
Cuprous oxide	Cu_2O	Powdery mildews
Basic copper sulfate	$CuSO_4 \cdot Cu(OH)_2 \cdot H_2O$	Seed treatment and preparation of Bordeaux mixture
Cupric carbonate	$Cu(OH)_2 \cdot CuCO_3$	Many fungal diseases

relatively insoluble or "fixed" copper fungicides, which release only very low levels of copper, adequate for fungicidal activity, but not enough to affect the host plant.

Copper compounds are not easily washed from leaves by rain, since they are relatively insoluble in water, and thus give longer protection against disease than do most of the organic materials. They are relatively safe to use and require no special precautions during spraying. Despite the fact that copper is an essential element for plants, there is some danger in an accumulation of copper in agricultural soils resulting from frequent and prolonged use. In fact, certain citrus growers in Florida have experienced a serious problem of copper toxicity after using fixed copper for disease control.

The currently accepted theory for the mode of action of copper's fungistatic action is its nonspecific denaturation of protein. The Cu^{++} ion reacts with enzymes having reactive sulfhydryl groups that would explain its toxicity to all forms of plant life, but especially its toxicity to the vulnerable fungal cells.

Mercury

The inorganic mercurial fungicides are probably the most toxic of the fungicides. Mercury's fungicidal properties and toxicity to animals are due in part to the degree of association of divalent mercury ions, which are toxic to all forms of life. As a result, no mercury residues are permitted in foods or feed.

Over the past 30 years, a host of organic mercury compounds was developed that has been replaced by other organic fungicides. Ceresan is typical of those used as seed treatments, whereas phenylmercury acetate (PMA) was useful for turf diseases, as a seed treatment, and as dormant sprays for fruit trees. The mode of action for the mercurials is the nonselective or nonspecific inhibition of enzymes, especially those containing iron and sulfhydryl sites.

Quite recently, both organic and inorganic mercurial fungicides, with one or two exceptions have been banned from home and agricultural use by the Environmental Protection Agency. The basis for this decision hinged on their toxicity to warm-blooded animals and accumulation of mercury in the environment.

THE ORGANIC FUNGICIDES

Many synthetic sulfur and other organic fungicides have been developed over the past 30 years to replace the more harsh, less selective inorganic materials. Most of them have had no measurable buildup effect on the environment after many years of use. The first of the organic sulfur fungicides was discovered in 1931; this fungicide, thiram, was followed by many others. Then came other new classes, the dithiocarbamates and dicarboximides (zineb and captan) introduced in 1943 and 1949, respectively. Since then, the organic synthesists have literally left the barn doors open, with now more than 125 fungicides of all classes in use and in various stages of development.

The newer organic fungicides possess several outstanding qualities. They are extremely efficient—that is, smaller quantities are required than of those used in the past; they usually last longer; and they are safer for crops, animals, and the environment. Most of the newer fungicides also have very low phytotoxicity, many being at least ten times safer than the copper materials. And most of them are readily degraded by soil microorganisms, thus preventing their accumulation in soils.

Dithiocarbamates

Among the dithiocarbamates we find the "old reliables." In this group are thiram, maneb, ferbam, ziram, Vapam® (SMDC), and zineb, all developed in the early 1930s and 1940s. Such fungicides probably have greater popularity and use than all other fungicides combined, including the home gardens. Except for systemic action, they are employed collectively in every use known for fungicides. The dithiocarbamates probably act by being metabolized to the isothiocyanate radical (—N=C=S), which inactivates the sulfhydryl groups in amino acids contained within the individual pathogen cells.

Let's turn now to an important function in plants that may explain the mode of action for dithiocarbamates as fungicides, namely chelation. Some of the metals required by the higher plants and fungi in trace amounts assist enzymes in conducting their routine duties of metabolism. The metal may be active in this role as a chelate with the biological component. A chelate is an organic ring structure composed of the metal atom linked into the ring by nitrogen, oxygen, or sulfur.

CERESAN

$CH_3OCH_2CH_2HgCl$

2-methoxyethylmercuric chloride

PMA

phenylmercury acetate

MANEB

manganese ethylenebisdithiocarbamate

FERBAM

ferric dimethyldithiocarbamate

ZINEB

zinc ethylenebisdithiocarbamate

Particularly as they involve enzymes, chelates are powerful and essential entities in the metabolic processes of plants. One of the generally accepted theories to explain the fungicidal activity of copper, mercury, cadmium, and other heavy metals is the formation of chelates within the fungal cells. The chelates, in turn, disrupt protein synthesis and metabolism. And since the most critical proteins of cells are enzymes, the metals required in trace amounts appearing in abundance or excessive quantities are equivalent to the introduction of potent poisons in the cells.

If the chelation theory is acceptable, the mode of action for the heavy metal fungicides, both organic and inorganic, and for the formation of isothiocyanates from the dithiocarbamate molecules is explained, and the potency of the heavy-metal dithiocarbamates becomes evident.

Also, certain fungicides are in themselves chelating agents. They attach to the scarce metal components, such as Fe, Mg, and Zn, within the cells, literally robbing cells of essential materials.

In summary, chelation of heavy metals plays an important role in both the life and death of cells.

Thiazoles

TERRAZOLE®

5-ethoxy-3-trichloromethyl-1,2,4-thiadiazole

The thiazoles, a class of compounds that offers a surprising chemical disposition, contains among others, Terrazole®. The five-membered ring of the thiazoles is cleaved rather quickly under soil conditions to form either the fungicidal isothiocyanate ($-N=C=S$) or a dithiocarbamate, depending on the structure of the parent molecule. Terrazole® is used only as a soil fungicide and, as such, is exposed to the ring-cleavage just mentioned. The probable mode of action is similar to that of the dithiocarbamates.

Triazines

ANILAZINE

2,4-dichloro-6-(o-chloroanilino)-s-triazine

The triazine structure, seen frequently in herbicides, is found only once in the fungicides. Anilazine was introduced in 1955 and has received wide use for control of potato and tomato leaf spots and turf-grass diseases.

Substituted Aromatics

The substituted aromatics belong in a somewhat arbitrary classification assigned to the simple benzene derivatives that possess long-recognized fungicidal properties.

Hexachlorobenzene, introduced in 1945, is used as a seed treatment and as a soil treatment to control stinking smut of wheat. Pentachlorophenol (PCP) has been used since 1936 as a wood preservative, as a seed treatment, and, as pointed out earlier, as an herbicide. Pentachloronitrobenzene (PCNB) was introduced in the 1930s as a fungicide for seed treatment and selected foliage applications. It is also used as a soil treatment to control the pathogens of certain damping-off diseases of seedlings. Chlorothalonil is a very useful, broad-spectrum foliage-protectant fungicide made available in 1964. And chloroneb, developed in 1965, is heavily used for cotton seedling and turf diseases.

HEXACHLOROBENZENE

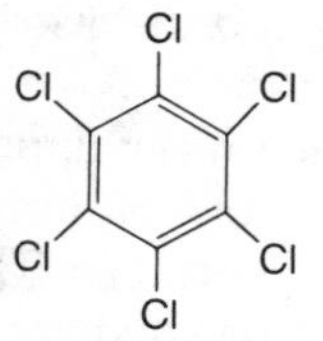

1,2,3,4,5,6-hexachlorobenzene

PCNB

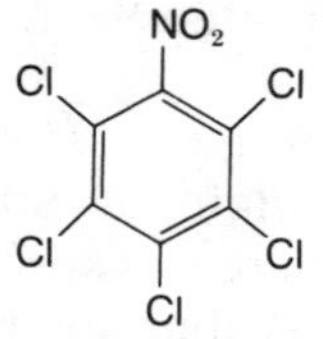

pentachloronitrobenzene

CHLORONEB

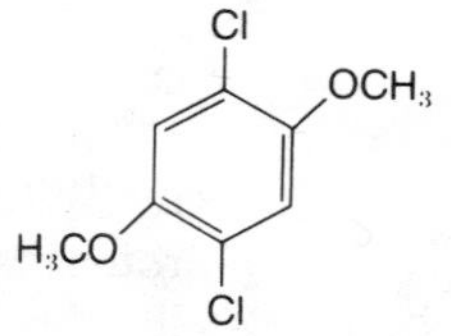

1,4-dichloro-2,5-dimethoxybenzene

PCP

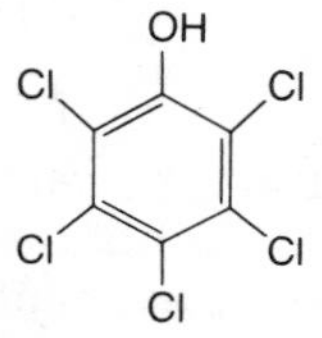

pentachlorophenol

CHLOROTHALONIL

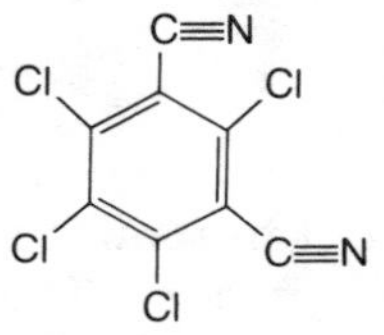

tetrachloroisophthalonitrile

CAPTAN

N-(trichloromethylthio)-4-cyclohexene-
1,2-dicarboximide

Substituted aromatics are diverse in their modes of action. Being generally fungistatic, they reduce growth rates and sporulation of fungi, probably by combining with —NH$_2$ or —SH groups of essential metabolic compounds.

Dicarboximides

Three extremely useful foliage protectant fungicides belong to this group: Captan appeared in 1949 and is undoubtedly the most heavily used fungicide around the home of all classes; folpet appeared in 1962; and captabol (Difolatan®) appeared in 1961. They are used primarily as foliage dusts and sprays on fruits, vegetables, and ornamentals.

The dicarboximides are some of the safest of all pesticides available, and are recommended for lawn and garden use, as seed treatments, and as protectants for mildews, late blight, and other diseases. Remember the garden center adage, "When in doubt, use captan."

Many compounds containing the —SCCl$_3$ moiety are fungitoxic, a fact that indicates that group as a toxophore. Fungitoxicity of the dicarboximides is apparently nonspecific and is not a result of a single mode of action. The dicarboximides' lethal effect on disease organisms is probably due to the inhibition of the synthesis of amino compounds and enzymes containing the sulfhydryl (—SH) radical.

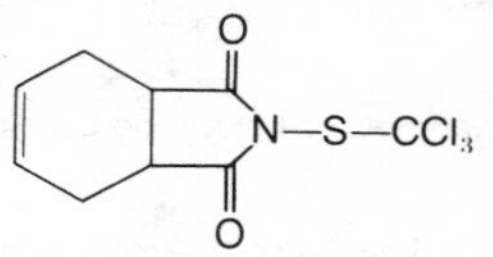

FOLPET

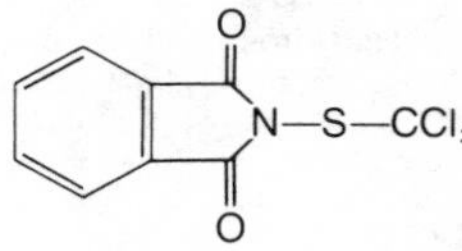

N-trichloromethylthiophthalimide

CAPTAFOL (Difolatan®)

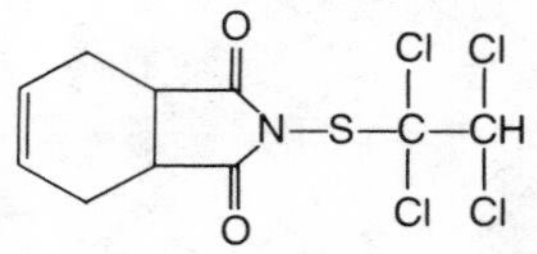

tetrachloroethylmercaptocyclohexenedicarboximide

Systemic Fungicides

It has only been in recent years that successful systemic fungicides have been marketed, and very few are yet available. Systemics are absorbed by the plant and carried by translocation through the cuticle and across leaves to the growing points. Most systemic fungicides have eradicant properties that stop the progress of existing infections. They are therapeutic in that they can be used to cure plant diseases. A few of the systemics can be applied as soil treatments and are slowly absorbed through the roots to give prolonged disease control.

These systemics offer much better control of diseases than is possible with a protectant fungicide that requires uniform application and remains essentially where it is sprayed onto the plant surfaces. There is, however, some redistribution of protective fungicidal residues on the surfaces of sprayed or dusted plants, giving them longer residual activity than would be expected.

Those systemics currently in commercial use will be mentioned in their chronological order of appearance. None are available for around-the-home use.

Oxathiins. The two systemic fungicides in this group, carboxin and oxycarboxin, both introduced in 1966, were the first of the systemics to succeed in practice. They are used as seed treatments for the cereal crops, particularly those affected by embryo-infecting smut fungi, and they have potential for other uses. They are selectively toxic to the smuts, rusts, and to *Rhizoctonia (Thanatephorus)*, organisms belonging to the Basidiomycetes. The apparent mode of action of the oxathiins begins with their selective concentration in the fungal cells, followed by the inhibition of succinic dehydrogenase, an important enzyme to respiration in the mitochondrial systems.

Benzimidazoles. The benzimidazoles, represented by benomyl and thiabendazole (TBZ), were introduced in 1968 and have received wide acceptance as systemic fungicides against a broad spectrum of diseases. Benomyl has the widest spectrum of fungitoxic activity of all the newer systemics, including the *Sclerotinia*, *Botrytis*, and *Rhizoctonia* species, and the powdery mildews and apple scab. Thiabendazole has a similar spectrum of activity to that of benomyl. Introduced in 1969, thiophanate, although not a benzimidazole in its original structure, is converted to that group by the host plant and the fungus through their metabolism. Thiophanate has a fungitoxicity similar to that of benomyl. All three compounds have been used in foliar applications, seed treatment, dipping of fruit or roots, and soil application. Their mode of action appears to be the induction of abnormalities in spore germination, cellular multiplication, and growth, as a result of their interference in the synthesis of that vital nucleic material, deoxyribonucleic acid (DNA).

The systemic fungicides cover susceptible foliage and flower parts more efficiently than protectant fungicides because of their ability to translocate through the cuticle and across leaves. They bring into

CARBOXIN

2,3-dihydro-5-carboxanilido-6-methyl-1,4-oxathiin

OXYCARBOXIN

2,3-dihydro-5-carboxanilido-6-methyl-
1,4-oxathiin-4,4-dioxide

BENOMYL

methyl 1-(butylcarbamoyl)-2-benzimidazolecarbamate

THIABENDAZOLE

2-(4′-thiazoyl) benzimidazole

THIOPHANATE

1,2-bis(3-ethoxycarbonyl-2-thioureido)benzene

play the perfect method of disease control by attacking the pathogen at its site of entry or activity, and they reduce the risk of contaminating the environment by frequent broadcast fungicidal treatments. Undoubtedly, as newer and more selective systemic molecules are synthesized, they will gradually replace the protectants that comprise the bulk of our fungicidal arsenal.

Disease Resistance to Fungicides

Resistance to chemicals other than the heavy metals occurs commonly in fungal and, on rare occasions in bacterial, plant disease pathogens. Several growing seasons after a new fungicide appears, it becomes noticeably less effective against a particular disease. As our fungicides become more specific for selected diseases, we can expect the pathogens to become resistant. This can be attributed to the singular mode of action of a particular fungicide that disrupts only one genetically controlled process in the metabolism of the pathogen. The result is that resistant populations appear suddenly, either by selection of resistant individuals in a population or by a single-gene mutation. Generally, the more specific the site and mode of fungicidal action, the greater the likelihood for a pathogen to develop a tolerance to that chemical.

Fumigants

As with the insecticides, there are several highly volatile, small-molecule fungicides that have fumigant action. They are unrelated chemically but are handled similarly and are dealt with as a single class in this book. Chloropicrin was mentioned before, as a warning agent in grain fumigants, but it is also an ideal fumigant itself: It controls fungi, insects, nematodes, and weed seeds in the soil. Methyl bromide, also listed as a fumigant insecticide, is equally effective against fungi, nematodes, and weeds. Methylisothiocyanate (MIT) is closely related to the dithiocarbamates and has a similar mode of action against fungi, nematodes, and weeds. SMDC was listed, but not illustrated, in the dithiocarbamate fungicides, where it belongs chemically. SMDC decomposes in the soil to yield methylisothiocyanate. None of these are available to the homeowner.

CHLOROPICRIN

trichloronitromethane

METHYL BROMIDE

CH_3Br

bromomethane

MIT

$CH_3—N=C=S$

methylisothiocyanate

Antibiotics

Antibiotics, as you probably know them, are those such as penicillin, tetracycline, and chloramphenicol, used medically against human bacterial diseases. These are not involved in our present study. But we should note in passing that the oxytetracyclines are therapeutic against some of those mysterious mycoplasmalike diseases. As in the case of the medically important antibiotics, the antibiotic fungicides are substances produced by microorganisms, which in very dilute concentrates inhibit growth and even destroy other microorganisms. To date, several hundred antibiotics have been reported to have fungicidal activity, and the chemical structures of about half of these are already known.

The largest source of antifungal antibiotics is the actinomycetes, a group of the lower plants. Within the actinomycetes is one amazing species, *Streptomyces griseus*, from which we obtain not one, but two antibiotics, streptomycin and cycloheximide.

Streptomycin is used as dust, spray, and seed treatment for the control mostly of bacterial diseases such as blight on apples and pears, soft rot on leafy vegetables, and some seedling diseases. It is also effective against a few fungal diseases.

The mode of action of streptomycin is not clearly understood, but it probably interferes in the synthesis of proteins. Despite the evidence of antibiotic-resistant strains, streptomycin has a place in the control of some bacterial diseases, and the tetracyclines may well play an important part in controlling some mycoplasmalike diseases of plants. Because of the scientific principles behind its development, resistance to antibiotics by both bacteria and mycoplasmas is inevitable.

STREPTOMYCIN

2,4-diguanidino-3,5,6-trihydroxycyclohexyl-5-deoxy-
2-*O*-(2-deoxy-2-methylamino-α-glucopyranosyl)-
3-formyl pentofuranoside

Cycloheximide is a smaller, less complicated antibiotic, about which more is understood. First, cycloheximide is toxic to a wide range of organisms, including yeasts, filament-forming fungi, algae, protozoa, higher plants, and especially mammals (see Table 10). Surprisingly, it is inactive against bacteria. As a matter for speculation, this may result from failure of bacteria to absorb it. Cycloheximide causes growth inhibition in yeasts and filament-forming fungi by inhibiting protein and RNA synthesis. Then, with a fair amount of confidence, we can say that both streptomycin and cycloheximide act by inhibiting the synthesis of nucleic acids.

Cycloheximide was introduced as a fungicide in 1949 and has since become popular in the control of powdery mildew, rusts, turf diseases, and certain blights. It is best known commercially under the name Actidione®. Because of its high acute toxicity, it cannot be purchased for home and garden use.

Dinitrophenols

Because of their previous appearances as insecticide-ovicides and herbicides, the dinitrophenols should be familiar by now. Their mode of action as fungicides is the same, uncoupling oxidative phosphorylation in cells with an attendant upset of the energy systems within the cells. This statement pertaining to their mode of action holds true in all cases.

One dinitrophenol, dinocap (Arathane®, Karathane®), has been used since the late 1930s, both as an acaricide and for powdery mildew on a number of fruit and vegetable crops. Dinocap undoubtedly acts in the vapor phase, since it is quite effective against powdery mildews whose spores germinate in the absence of water. This is a popular home fungicide.

Quinones

The quinones are a fascinating chemical group, because there are practically unending numbers of molecules that are potential fungicides. The one example reviewed briefly here is the most popular of that group, dichlone. It is used on a number of fruit and vegetable crops and for treatment of ponds, to control blue-green algae. Dichlone affects respiration in many fungi and acts by attaching to the —SH groups in enzymes, thus inhibiting their action and indirectly uncoupling oxidative phosphorylation.

Organotins

The organotins were first introduced in the mid-1960s, ten years after their fungicidal properties had been discovered. In general, the trialkyl derivatives are highly fungicidal, but also phytotoxic. The triaryl (three-ring) compounds are suitable for protective use, and also have acaricidal properties. Although it is not yet proved, the trisubstituted tin compounds probably block oxidative phosphorylation, as mentioned in the section on insecticides.

Aliphatic Nitrogen Compounds

Dodine, a fungicide introduced in the mid-1950s, has proved effective in controlling certain diseases such as apple and pear scab and cherry leaf spot. It has disease specificity and slight systemic qualities. Its mode of action is not totally clear, but it is taken up rapidly by fungal cells, causing leakage in these cells, possibly by alteration in membrane permeability. The guanidine nucleus of dodine is also known to inhibit the synthesis of RNA.

CYCLOHEXIMIDE

β[2-(3,5-dimethyl-2-oxocyclohexyl)-2-hydroxyethyl]-glutarimide

DINOCAP

2,4-dinitro-6-(2-octyl)phenyl crotonate

DICHLONE

2,3-dichloro-1,4-naphthoquinone

FENTIN HYDROXIDE

triphenyltinhydroxide

DODINE

n-dodecylguanidine acetate

CHAPTER 17

Somewhere, behind Space and Time,
Is wetter water, slimier slime!

Rupert Brooke

Algicides

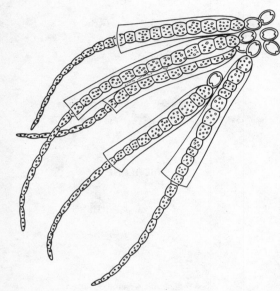

Algae are a group of simple freshwater and marine plants, ranging from single-celled organisms to green pond scums and very long seaweeds (kelps). The algicides are chemicals intended for the control of algae, especially in water that is stored or is being used industrially. (Your swimming pool qualifies!)

INORGANIC CHLORINES

Swimming pools are a good place to begin. That strong smell of chlorine issuing from the YMCA or other public pool is probably released from the ever-popular calcium hypochlorite or chloride of lime, not only a good algicide but an excellent disinfectant as well. It contains 70 percent available chlorine and is the source of most bottled laundry bleaches (7 to 8 percent $CaOCl_2$).

There are several chlorine-based inorganic salts available for pool chlorination. Some of the more popular are:

1. Calcium hypochlorite, $Ca(ClO)_2$.
2. Sodium hypochlorite, $NaClO$.
3. Lithium hypochlorite, $LiClO$.
4. Sodium chlorite, $NaClO_2$.

COPPER COMPOUNDS

Any of the copper-containing algicides would be equally effective, and longer lasting, than the chlorine-based materials. However, the copper content may eventually become phytotoxic to grass or other herbaceous plants surrounding the pool that are splashed or drenched occasionally. This principle applies to other algicides of greater potency.

The organic copper complexes are not for the swimming pool, but rather for industrial, public water systems and agriculture. Most commonly used among these are the copper-triethanolamine materials. Generally they can be used as surface sprays for filamentous and planktonic forms of algae in potable water reservoirs; irrigation water storage and supply systems; farm, fish, and fire ponds; and lake and fish hatcheries. The treated water can be used immediately for its intended purpose. The same generalities apply also to another organic copper compound, copper ethylenediamine complex.

Another of the copper group is the old standby, copper sulfate pentahydrate, whose older names are *bluestone* and *blue vitriol*. In all instances, it is the copper ion in solution that is lethal to the simple single-cell and filamentous algae. The control principle is the same as that involved in the copper-based fungicides mentioned earlier.

QUATERNARY AMMONIUM HALIDES

A host of algicides belong to these quaternary ammonium (QA) compounds. The QA's are characterized by having a chlorine or bromine ion on one end and a nitrogen atom with four carbon-nitrogen bonds. At least one of the carbons is the extension of an 8- to 18-carbon chain (R, in the formula). The straight carbon chain is derived from fatty acids similar to those found in vegetable oil. A generalized formula follows:

$$R-\underset{\underset{CH_3}{|}}{\overset{\overset{CH_3}{|}}{N}}-CH_2Cl$$

They are general-purpose antiseptics, germicides, and disinfectants, ideal for algae control in the greenhouse—as pot dips and wall, bench, and floor sprays—and in swimming pool and recirculation water systems. Algae control lasts up to several months.

The alkyldimethylbenzylammonium chlorides can be used to give long-lasting control of algae and bacteria in swimming pools, cooling systems, air-conditioning systems and glass houses. They compose the bulk of the quarternary ammonium halides. As they are toxic to fish, they cannot be used in ponds, lakes, or streams.

ALKYLDIMETHYLBENZYLAMMONIUM CHLORIDE

MISCELLANEOUS ORGANIC COMPOUNDS

To be used only by professionals is acrolein, a highly volatile organic molecule with tremendous physiological activity, both against aquatic plants and humans. It is extremely hazardous to work with because of its lachrymatory (violent action on eyes) effect and because skin burns result from contact.

Acrolein is an extremely useful aquatic herbicide and algicide but requires application by licensed operators because of its frightening tear-gas effect. Exposed plants disintegrate within a few hours and float downstream. It is a general plant toxicant, destroying plant cell membranes and reacting with various enzyme systems. By using spot treatment in lakes, the fish population can be saved, an example of environmental protection by spot application.

ACROLEIN

$$CH_2=CH-CHO$$

2-propenal

TRIPHENYLTIN ACETATE

Triphenyltin acetate (Brestan®) is another one of the multipurpose pesticides. In this instance, it can act as a fungicide, algicide, or molluscicide.

Dimanin C (sodium dichloroisocyanurate) also gives long-lasting control of algae in swimming pools and can be used as a disinfectant.

Simazine, a herbicide introduced earlier, has just recently been registered by EPA as an algicide under the name of Aquazine®. It performs well at a concentration of 1.0 ppm against a broad spectrum of algae and may be used in ponds containing fish.

Dichlone, an agricultural fungicide, is also registered as an algicide for use against blue-green algae in lakes and ponds.

SODIUM DICHLOROISOCYANURATE

POTASSIUM DICHLOROISOCYANURATE

SIMAZINE

2-chloro-4,6-bis(ethylamino)-s-triazine

DICHLONE

2,3-dichloro-1,4-naphthoquinone

CHAPTER 18

"Gentlemen, it is the microbes that will have the last word."

Louis Pasteur

Disinfectants

There are an almost limitless number of chemical agents for controlling microorganisms, and new ones appear on the market regularly. These are the disinfectants. A common problem confronting persons who must utilize disinfectants or antiseptics is which one to select and how to use it. There is no single ideal or all-purpose disinfectant, thus the compound to choose is the one that will kill the organisms present in the shortest time, with no damage to the contaminated substrate.

In the "Pesticides Bible," the EPA Compendium of Registered Pesticides, Volume V, devoted entirely to disinfectants (all microbiological control producers, e.g., sterilizers, disinfectants, bacteriostats, sanitizers, viricides, and microbiocides), there are listed approximately 440 active ingredients in disinfectant products. There are 23 major use categories for disinfectants, and these are further divided into 200 use sites. The major use categories are as follows:

> Additives, antimicrobial
> Animal (for food use) quarters
> Animal (for nonfood use) quarters
> Aquatic sites
> Barber and beauty shops
> Beverage processing plants
> Carpets
> Chemical lavatory holding tanks
> Dairy processing plants
> Dry cleaning
> Eating establishments (restaurants, bars, or taverns)
> Food-processing plants
> Funeral homes, mortuaries, and morgues
> Hospitals and related institutions
> Laundry
> Maintenance, commercial and industrial
> Maintenance, household
> Milk samples
> Oil recovery process and oil well drilling
> Preservatives, food and feed
> Preservatives, industrial
> Specialty sites
> Toilet bowls

This list should give the reader some idea as to the number of disinfectants available and their range of uses.

The history of the development of disinfectants is a fascinating documentation of the misunderstanding of the importance of microorganisms. During the early 1800s, microorganisms were considered to be biological nonentities. Infections were believed to be caused by magical powers in the air, or by an imbalance of body fluids. Certainly contaminated hands were not involved. It took the dogged tenacity of Louis Pasteur to demonstrate to the world that microbes could not only ferment fruit juice to wine but could also cause the wine to spoil, before the idea of the possible role of microorganisms in disease could evolve.

The next steps were made in the 1860s by an English surgeon, Dr. Joseph Lister, who first developed antiseptic surgery with the use of heat-sterilized instruments and the application of carbolic acid (phenol) to wounds by means of soaked dressings. In 1881, the German bacteriologist, Robert Koch, evaluated 70 different chemicals for use in disinfection and antisepsis. Among these were various phenols and mercuric chloride ($HgCl_2$). From this point on, progress in the science of destroying our microorganismic enemies (for the most part) has been overwhelming.

Before we begin the study of the chemicals referred to as *disinfectants,* we must make a quick distinction between two confusing words, *antisepsis* and *sanitation*. Antisepsis is the disinfection of skin and mucous membranes, while sanitation is the disinfection of inanimate surfaces. Consequently, much more severe treatment can be used for sanitation than for antisepsis. In this chapter, we will direct our presentation to the use of disinfectants used in sanitation.

THE PHENOLS

Because Dr. Lister used phenol in the 1860s as a germicide in the operating room, phenol can probably be considered the oldest recognized disinfectant. When greatly diluted, its deadly effect is due to protein precipitation. It is used as the standard for the comparison of the activities of other disinfectants, expressed in terms of phenol coefficients. Phenol and the cresols have very distinct odors, which change little with modification of their chemical structures. The addition of chlorine or a short-chain organic compound increases the activity of the phenols. For instance, hexachlorophene is one of the most useful of the phenol derivatives but is in a state of some disrepute at this writing. After much debate, it was removed from all over-the-counter items, including soaps and shampoos.

THE HALOGENS

These are compounds containing chlorine, iodine, bromine, or fluorine, both organic and inorganic. Generally, the inorganic halogens are deadly to all living cells.

Phenol and Several Cresol and Hexachlorophene Structures

PHENOL

ORTHOCRESOL

META CRESOL

PARA CRESOL

CRESYLACETATE

HEXACHLOROPHENE

Chlorine

Chlorine in its many forms was first used as a deodorant and later as a disinfecting agent. It is a standard treatment for drinking water in all communities of the United States. Hypochlorites are those most commonly used in disinfecting and deodorizing procedures because they are relatively safe to handle, colorless, good bleaches, and do not stain. Several organic chlorine derivatives are used for the disinfection of water, particularly for campers, hikers, and the military. The most common of these are halazone and succinchlorimide.

Chlorine is the dominant element in disinfectants, in that roughly 25 percent of those registered with EPA contain one or more atoms of this important halogen.

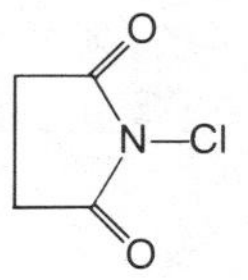

Hypochlorites

Calcium hypochlorite and sodium hypochlorite are popular compounds widely used both domestically and industrially. They are available as powders or liquid solutions and in varying concentrations depending on the use. Products containing 5 to 70 percent calcium hypochlorite are used for sanitizing dairy equipment and eating utensils in restaurants. Solutions of sodium hypochlorite are used as a household disinfectant; higher concentrations of 5 to 12 percent are also used as household bleaches and disinfectants and for use as sanitizing agents in dairy and food-processing establishments.

Chloramines

This is another category of chlorine compounds used as disinfectants, or sanitizing agents. Chemically they are characterized by having one or more hydrogen atoms of an amino group of a compound replaced with chlorine. The simplest of these is monochloramine, NH_2Cl. Chloramine-T and azochloramide, two of several germicidal compounds in this general group, have more complex chemical structures. One of the advantages of the chloramines over the hypochlorites is their stability and prolonged chlorine release.

The germicidal action of chlorine and its compounds is produced by the formation of hypochlorous acid when free chlorine reacts with water:

$$Cl_2 + H_2O \rightarrow HCl + HClO \text{ (hypochlorous acid)}$$

Similarly, hypochlorites and chloramines undergo hydrolysis, forming hypochlorous acid. The hypochlorous acid formed in each instance is further decomposed, releasing oxygen:

$$HClO \rightarrow HCl + O \text{ (formed from chlorine, hypochlorites, and chloramines)}$$

The oxygen released in this reaction (nascent oxygen) is a strong

HALAZONE

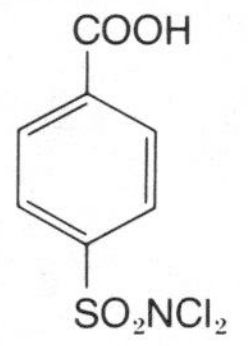

p-sulfone dichloramidobenzoic acid

SUCCINCHLORIMIDE

CHLORAMINE-T

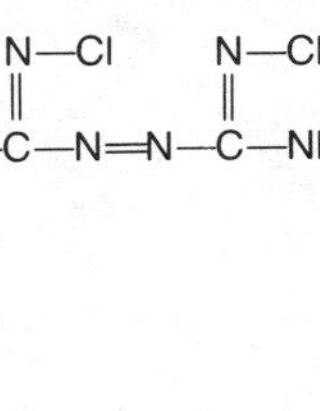

AZOCHLORAMIDE

oxidizing agent, and microorganisms are destroyed by its action on cellular constituents. The killing of cells by chlorine and its compounds is also in part caused by the direct combination of chlorine with proteins of the cell membranes and enzymes.

Iodine

Water or alcohol solutions of iodine are highly antiseptic and have been used for decades before surgical procedures. Several metallic salts, such as sodium and potassium iodide, are registered as disinfectants, but the number of compounds containing iodine nowhere approaches those containing chlorine.

Fluorine

Because of its extreme reactivity, and lack of easy-to-handle characteristics, fluorine in combination with other elements is found in only a handful of disinfectants. The same is true for bromine.

HEAVY METALS

The heavy metals, either alone or in certain chemical compounds, usually exhibit their deadly effect by precipitating proteins or by reacting with enzymes or other essential cellular components. The same effects were discussed in the chapter on fungicides, which also mentioned some of the heavy metals. In fact some of the fungicides are disinfectants. The heavy metals in common use are mercury, silver, arsenic, zinc, and copper.

Mercury

Mercury compounds are historic, in that virtually every registration of both the inorganic and organic mercurial fungicides and disinfectants has been cancelled by the EPA following the "mercury-in-the-environment" crisis. Prominent among the historic compounds was mercuric bichloride, $HgCl_2$, which at one time or another was used in every conceivable situation where a disinfectant was needed.

Copper

Copper is much more effective against algae and molds than bacteria and is used in the forms of copper sulfate and copper ethylenediaminetetra-acetate. A concentration of 2 ppm in water is sufficient to prevent algal growth and is used in swimming pools and open-water reservoirs.

Silver

Of the silver compounds, only silver fluoride and silver nitrate have retained registrations and are used only as antiseptics.

Arsenic

Arsenic achieved fame as the first-known treatment for syphilis and still finds some use in the treatment of protozoan infections, but is not used in sanitation procedures.

Zinc

Fungi are particularly vulnerable to the several zinc compounds, two of which are common household garden fungicides, zineb and ziram.

THE DETERGENTS

The detergents are organic compounds that have two ends or poles. One is hydrophilic and mixes well with water. The other is hydrophobic and does not mix well with water. As a result, the compounds orient themselves on the surfaces of objects with their hydrophilic poles toward the water. Basically, these may be classed as *ionic* or *nonionic* detergents. The ionic are either anionic (negatively charged) or cationic (positively charged). The anionic detergents are only mildly bactericidal. The cationic materials, which are the quaternary ammonium compounds, are extremely bactericidal, especially for *Staphylococcus*, but do not affect spores. Hard water, containing calcium or magnesium ions, will interfere with their action, and they also rust metal objects. Even with these disadvantages, the cationic detergents are among the most widely used disinfecting chemicals, since they are easily handled and are not irritating to the skin in concentrations ordinarily used.

Structure of a quaternary ammonium compound shown in relation to the structure of ammonium chloride

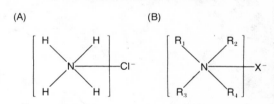

(A) Ammonium chloride; (B) the general structure of a quaternary ammonium compound—R_1, R_2, R_3, and R_4 are carbon-containing groups, and the X^- is a negatively charged ion, such as Cl^- and Br^-.

Detergent-Disinfectant Molecules

NONIONIC DETERGENT

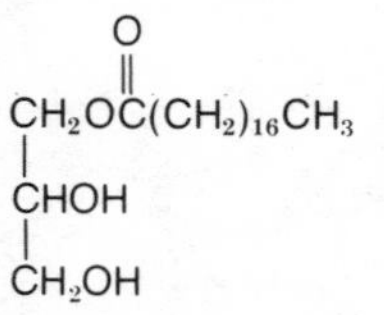

$$CH_2OC(CH_2)_{16}CH_3$$
$$|$$
$$CHOH$$
$$|$$
$$CH_2OH$$

Stearic acid monoglyceride

ANIONIC DETERGENT

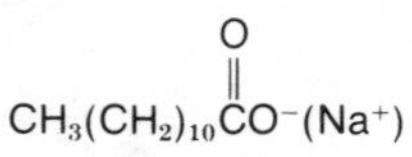

$$CH_3(CH_2)_{10}CO^-(Na^+)$$

Sodium laurate

CATIONIC DETERGENT

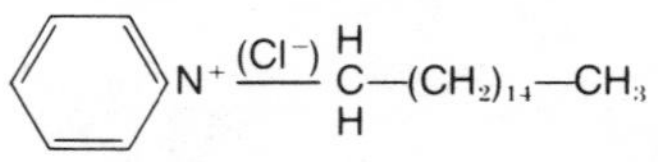

Cetylpyridinium chloride

THE ALDEHYDES

Combinations of formaldehyde and alcohol are outstanding steriliz-ing agents, with the exception of the residue remaining after its use. A related compound, glutaraldehyde, in solution is as effective as formaldehyde. Most organisms are killed in 5 minutes' exposure to glutaraldehyde, while bacterial spores succumb in 3 to 12 hours.

FORMALDEHYDE

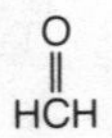

$$\overset{\displaystyle O}{\underset{\displaystyle HCH}{\|}}$$

GLUTARALDEHYDE

$$\overset{\displaystyle O}{\|}\qquad\overset{\displaystyle O}{\|}$$
$$HCCH_2CH_2CH_2CH$$

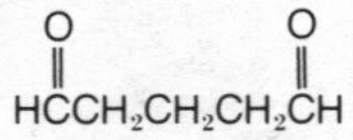

There are other classes of disinfectants, including a number of dyes, acids and alkalis, alcohols, peroxides, and fumigants (ethylene oxide and methyl bromide), which are very effective under certain conditions. However, due to the breadth, rather than the depth of this book, they are only mentioned in passing.

In summary, remember that no single chemical antimicrobial agent is best or ideal for any and all purposes. This is not surprising, in view of the variety of conditions under which agents may be used, the differences in modes of action, and the many types of microbial cells to be destroyed.

PART VI

LEGALITY AND HAZARDS
OF PESTICIDE USE

CHAPTER 19

The Toxicity and Hazards of Pesticides

DANGER

POISON

Dosage alone determines poisoning.
Paracelsus (1564)

Let's get it straight at the outset: There is a marked distinction between *toxicity* and *hazard*. These two terms are not synonymous. The word *toxicity* refers to the inherent toxicity of a compound—in other words, to how toxic it is to animals under experimental conditions. The word *hazard* refers to the risk or danger of poisoning when a chemical is used or applied, sometimes referred to as *use hazard*. The factor with which the user of a pesticide is really concerned is the use hazard and not the inherent toxicity of the material. Hazard depends not only on toxicity but also on the chance of exposure to toxic amounts of the material.

The dictionary definition of the word *poison* is "any substance which, introduced into an organism in relatively small amounts, acts chemically upon the tissues to produce serious injury or death." One can immediately spot several flaws in this definition. The "relatively small amount" statement is open to wide interpretation. For instance, many chemical agents to which man is exposed regularly could be termed *poisons* under this definition. An oral dose of 400 milligrams per kilogram (mg/kg) of sodium chloride, ordinary table salt, will make a person violently ill. A standard aspirin tablet contains about 5 grains of aspirin, chemically known as *acetysalicylic acid*. A fatal dose of aspirin for humans is in the range of 75 to 225 grains or 15 to 45 tablets. Approximately 85 deaths occur every year (about one-third are children) as a result of overdoses of aspirin. To take a third example, let us consider nicotine. A fatal oral dose of this naturally occurring alkaloid for humans is about 50 milligrams (mg), approximately the amount of nicotine contained in two unfiltered cigarettes. However, in smoking most of the nicotine is decomposed by burning, and thus it is not absorbed by the smoker.

In each of these cases, humans are not exposed during ordinary use to the amounts of salt, aspirin, and nicotine that cause toxicity problems. Therefore, it is obvious that the *hazard* from normal exposure is very slight even though the compounds themselves would be toxic under other circumstances.

A more adequate definition for the term *poison* has been suggested: "A chemical substance which exerts an injurious effect in the majority of cases in which it comes into contact with living organisms during normal use." The compounds mentioned above would obviously be excluded by such a definition, and so would the majority of pesticides.

Pesticides, by necessity, are poisons, but the toxic hazards of different compounds vary greatly. As far as the possible risks associated with the use of pesticides are concerned, we can distinguish between two types: (1) acute poisoning, resulting from the handling and application of toxic materials; and (2) chronic risks from long-term exposure to small quantities of materials or from ingestion of them. The question of acute toxicity is obviously of paramount interest to people engaged in manufacturing and formulating pesticides and to those responsible for their application. Supposed chronic risks, however, are of much greater public interest, because of their potential effect on the consumer of agricultural products.

Fatal human poisoning by pesticides is uncommon in the United States and is due to accident, ignorance, suicide, or crime. Fatalities represent only a small fraction of all recorded cases of poisoning, as demonstrated by these recent United States statistics (Table 5). Note that in 1968, 2.8 percent of the deaths from accidental poisoning were from pesticides, while in 1974 this cause had dropped to less than 1.0 percent. When the children's part of these statistics are viewed, it becomes a much more serious matter (Table 6).

In 1968, 11 percent of all accidental poisoning deaths were children under 5 years of age, while in 1974 that figure had dropped to 3.4 percent, a remarkable improvement. Still another, but not so good, statistic is that, of the children under 5 poisoned in 1968, 11 percent again were from pesticides, while in 1974 that percentage was little changed, at 9.6 percent. More devastating is the statement that of all deaths attributed to accidental poisoning by pesticides in 1974, 37 percent were children under 5 years of age.

In looking at another type of data for 1973, of 117,589 reported accidental ingestions among children under 5 years of age, pesticides were responsible for only about one-half the number attributed to cosmetics or two-thirds of those assigned to aspirin.

Regardless of your views toward pesticides and their presumed hazard, they have a fairly decent safety record, and it is growing better each year, mainly through education and labeling of containers.

EFFECTS OF PESTICIDES ON HUMANS

Pesticides have been developed to kill unwanted organisms and are toxic materials that produce their effects by several different mechanisms. Under certain conditions, they may be toxic to humans, and an understanding of the basic principles of toxicity and the differences between toxicity and hazard is essential. As you already know, some pesticides are much more toxic than others, and severe illness may result when only a small amount of a certain chemical has been ingested, while with other compounds no serious effects would result even after ingesting large quantities. Some of the factors that

TABLE 5
Total deaths from accidental poisoning.

	1968	1974
Medicines	1,692	2,742
Alcohol	182	370
Cleaning and polishing agents	23	13
Disinfectants	9	8
Paints and varnishes	2	1
Petroleum products and other solvents	70	54
Pesticides, fertilizers, or plant foods	72 (2.8%)	35 (0.9%)
Heavy metals (and their fumes)	53	23
Corrosives and caustics	28	13
Noxious foodstuffs and poisonous plants	10	5
Other unspecified solid and liquid	442	752
Total	2,583	4,016

Source: Mortality Statistics—Special Reports, Accident Fatalities, Division of Vital Statistics, National Center for Health Statistics, U.S. Department of Health, Education and Welfare, Washington, D.C.

TABLE 6
Accidental ingestions among children under 5 years of age.

Type of substance	1970		1973	
	Number	Percent	Number	Percent
Medicines	35,189	49.6	52,113	44.3
Internal	29,923	42.2	42,215	35.9
Aspirin	9,610	13.6	7,763	6.6
Other	20,313	28.7	34,452	29.3
External	5,266	7.4	9,898	8.4
Cleaning and polishing agents	10,810	15.2	19,132	16.3
Petroleum products	3,254	4.6	4,974	4.2
Cosmetics	5,112	7.2	10,362	8.8
Pesticides	3,887	5.5	5,591	4.8
Gases and vapors	55	0.1	140	0.1
Plants	3,574	5.0	7,032	6.0
Turpentine, paints, etc.	4,006	5.7	6,988	5.9
Miscellaneous	4,585	6.5	10,517	9.0
Not specified	425	0.6	740	0.6
Total	70,897	100.0	117,589	100.0

Source: Individual case reports submitted to the National Clearinghouse for Poison Control centers; 1973, from 517 centers in 45 states; 1970 from 432 centers in 49 states. *Bulletin of the National Clearinghouse for Poison Control Centers,* U.S. Food and Drug Administration, Bureau of Drugs, U.S. Department of Health, Education and Welfare, May–June 1974, p. 2.

influence this are related to (1) the toxicity of the chemical, (2) the dose of the chemical, especially concentration, (3) length of exposure, and (4) the route of entry or absorption by the body.

In the early stages of developing a pesticide for further experiments and exploration, toxicity data are collected on the pure toxicant, as required by the Environmental Protection Agency. These tests are conducted on test animals that are easy to work with and whose physiology, in some instances, is like that of humans; for example, white mice, white rats, white rabbits, guinea pigs, and beagle dogs. For instance, intravenous tests are determined usually on mice and rats, whereas dermal tests are conducted on shaved rabbits and guinea pigs. Acute oral toxicity determinations are most commonly made in rats and dogs, with the test substance being introduced directly into the stomach by tube. Chronic studies are conducted on the same two species for extended periods, and the compound is usually incorporated in the animal's daily ration. Inhalation studies may involve any of the test animals, but rats, guinea pigs, and rabbits are most commonly used.

All of these procedures are necessary to determine the overall toxicity is defined by the LD_{50}, expressed as milligrams (mg) of toxicant per kilogram (kg) of body weight, the dose that kills 50 percent of eventually some microlevel portion of the pesticide may be permitted in food for humans as a residue, which is expressed in ppm.

Pesticide toxicologists use rather simple animal toxicity tests to rank pesticides according to their toxicity. Long before pesticides are registered with the Environmental Protection Agency and eventually released for public use, the manufacturer must declare the toxicity of their pesticide to the white rat under laboratory conditions. This toxicity is defined by the LD_{50}, expressed as milligrams (mg) of toxicant per kilogram (kg) of body weight, the dose that kills 50 percent of the test animals to which it is administered under experimental conditions.

The LD_{50} is measured in terms of oral (fed to, or placed directly in the stomachs of rats), dermal (applied to the skin of rats or rabbits), and respiratory toxicity (inhaled). Using two of these tests, oral and dermal, a toxicologic ranking is shown for the organophosphate and organochlorine insecticides in Figures 3 and 4. The materials on the top of the list are the most toxic and those at the bottom the least. The size of the dose is the most important single item in determining the safety of a given chemical, and actual statistics of human poisonings correlate reasonably well with these toxicity ratings.

ESTIMATION OF TOXICITY TO HUMANS

The dose, length of exposure, and route of absorption are the other important variables beside toxicity. The amount of pesticide required to kill a human being can be correlated with the LD_{50} of the material to rats in the laboratory. In Table 7, for example, the acute oral LD_{50}, expressed as mg/kg dose of the technical material, is translated into the amount needed to kill a 170-lb human. Dermal LD_{50}s are included for a better understanding of the relationship of expressed animal toxicity to human toxicity.

Generally speaking, oral ingestions are more toxic than respiratory inhalations, which are more toxic than dermal absorption. Additionally, there are physical and chemical differences between pesticides that make them more likely or less likely to produce poisoning. For instance, parathion changes to the more toxic metabolite paraoxon under certain conditions of humidity and temperature.

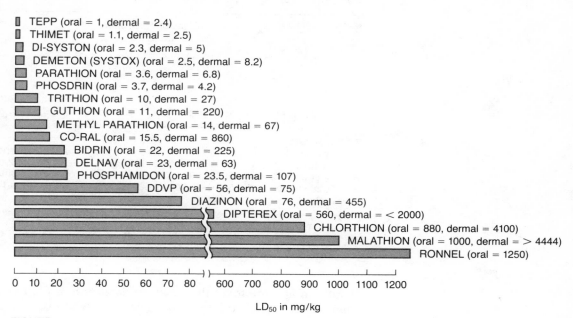

FIGURE 3
Acute oral and dermal toxicity values to rats for some organophosphate pesticides.
(*Source:* Unpublished chart prepared by the Bureau of Occupational Health, State
of California Department of Public Health. Reproduced by permission.)

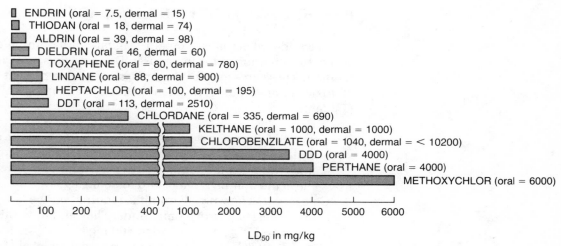

FIGURE 4
Acute oral and dermal toxicity values to rats for some chlorinated hydrocarbon
pesticides. (*Source:* Unpublished chart prepared by the Bureau of Occupational
Health, State of California Department of Public Health. Reproduced by
permission.)

TABLE 7
Combined tabulation of pesticide toxicity classes.

	Routes of absorption		
Toxicity rating	LD_{50} Single oral dose for rats, mg/kg	LD_{50} Single dermal dose for rabbits, mg/kg	Probable lethal oral dose for man
6—Supertoxic	<5	<20	A taste, a grain
5—Extremely toxic	5–50	20–200	A pinch, 1 teaspoon
4—Very toxic	50–500	200–1,000	1 teaspoon to 2 tablespoons
3—Moderately toxic	500–5,000	1,000–2,000	1 ounce to 1 pint
2—Slightly toxic	5,000–15,000	2,000–20,000	1 pint to 1 quart
1—Practically nontoxic	>15,000	>20,000	>1 quart

Source: Toxicity ratings modified from M. N. Gleason, R. E. Gosselin, and H. C. Hodge. 1976. *Clinical Toxicology of Commercial Products.* 4th ed. Williams and Wilkins Company, Baltimore, Md. p. 6.

Parathion is more toxic than methyl parathion to field workers; yet there is not a great difference in their oral toxicities. Workers' exposure is usually dermal, which explains why many more illnesses are reported in workers exposed to parathion than those exposed to methyl parathion. So we see that toxicity, route of absorption, dose, length of exposure, and the physical and chemical properties of the pesticide all contribute to its relative hazard. Hazard, then, is an expression of the potential of a pesticide to produce human poisoning.

PESTICIDE TOXICITY AND ITS LABEL

All pesticide labels must contain "signal words" in bold print, to attract the attention of the buyer/user: *Danger—Poison; Warning;* and *Caution.* These are significant words, since they represent a category of toxicity, and thus give an indication of their potential hazard (Table 8).

Category I. The signal words *Danger—Poison* and the skull and crossbones symbol are required on the labels for all *highly toxic* compounds. These pesticides all fall within the acute oral LD_{50} range of 0 to 50 mg/kg.

Category II. The word *Warning* is required on the labels for all *moderately toxic* compounds. They all fall within the acute oral LD_{50} range of 50 to 500 mg/kg.

Category III. The word *Caution* is required on labels for *slightly toxic* pesticides that fall within the LD_{50} range of 500 to 5000 mg/kg.

Category IV. The word *Caution* is required on labels for compounds having acute LD_{50}s greater than 5,000 mg/kg. However, unqualified claims for safety are not acceptable on any label, and all labels must bear the statement, "Keep Out Of Reach Of Children."

TABLE 8
EPA labeling toxicity categories by hazard indicator.

HAZARD INDICATORS	TOXICITY CATEGORIES			
	I	*II*	*III*	*IV*
Oral LD$_{50}$	Up to and including 50 mg/kg	From 50 thru 500 mg/kg	From 500 thru 5000 mg/kg	Greater than 5000 mg/kg
Inhalation LD$_{50}$	Up to and including 0.2 mg/liter	From 0.2 thru 2 mg/liter	From 2 thru 20 mg/liter	Greater than 20 mg/liter
Dermal LD$_{50}$	Up to and including 200 mg/kg	From 200 thru 2000	From 2000 thru 20,000	Greater than 20,000
Eye effects	Corrosive; corneal opacity not reversible within 7 days	Corneal opacity reversible within 7 days; irritation persisting for 7 days	No corneal opacity; irritation reversible within 7 days	No irritation
Skin effects	Corrosive	Severe irritation at 72 hours	Moderate irritation at 72 hours	Mild or slight irritation at 72 hours

Source: "EPA Pesticide Programs, Registration and Classification Procedures, Part II." *Federal Register 40:* 28279.

Table 9 shows the relative acute toxic hazards of many of the commonly used pesticides to spraymen. This is a valuable reference. Examples of insecticides, herbicides, and fungicides in the three label toxicity classifications are presented in Table 10.

With regard to the classifications of pesticides, their general toxicities, in decreasing order, would be insecticides > defoliants > desiccants > herbicides > fungicides. Within the most toxic class, the insecticides, the categories would fall in the following general order of their dermal hazards to humans: organophosphates > carbamates > cyclodienes > DDT-relatives > botanicals > activators or synergists > inorganics. There are usually exceptions in each category listed.

The formulations of pesticides, because of their varying kinds of diluents, would also have varying degrees of hazard to humans. Again, we must generalize: liquid pesticide > emulsifiable concentrate > oil solution > water emulsion > water solution > wettable powder/flowable (in suspension) > dust > granular.

In closing, just a word regarding the toxicity of the petroleum solvents used in formulating pesticides. Currently these are diesel fuel, deodorized kerosene, methanol, petroleum distillates, xylene, and toluene. Of these, only the xylene and toluene are aromatics, and they offer by far the greater dermal hazard.

TABLE 9
Estimated relative acute toxic hazards of pesticides to spraymen.[a]

Most dangerous	Dangerous
aldicarb, Temik® (C)[b]	aldrin (CH)
demeton, Systox® (OP)	carbophenothion, Trithion®
disulfoton, Di-Syston® (OP)	DDVP, dichlorvos (OP)
mercuric chloride (M)	dicrotophos, Bidrin® (OP)
methylmercury chloride (M)	dieldrin (CH)
mevinphos, Phosdrin® (OP)	dioxathion, Delnav® (OP)
parathion (OP)	DNOC (N)
phenylmercuric acetate (M)	DNOSBP (N)
phorate, Thimet® (OP)	endrin (CH)
tepp (OP)	EPN (OP)
thionazin, Zinophos® (OP)	ethion, Nialate® (OP)
	methyl parathion (OP)
	mexacarbate, Zectran® (C)
	monocrotophos, Azodrin® (OP)
	nicotine (M)
	Paraquat® (M)
	pentachlorophenol (M)
	phosphamidon, Dimecron® (OP)
	sodium arsenite (M)

Less dangerous[c]	Least dangerous
azinphosmethyl, Guthion® (OP)	captan (M)
BHC (CH)	carbaryl, Sevin® (C)
binapacryl, Morocide® (N)	carbofuran, Furadan® (C)
chlordane (CH)	chlorobenzilate (CH)
chlordimeform, Fundal®, Galecron® (F)	chlorpyrifos, Dursban® (OP)
coumaphos, Co-Ral® (OP)	2,4-D (CH)
crufomate, Ruelene® (OP)	DDD, TDE (CH)
diazinon (OP)	DDT (CH)
dicapthon (OP)	dicofol, Kelthane® (CH)
dichlofenthion, VC-13 (OP)	dinocap, Karathane® (N)
dichloroethyl ether (M)	diquat (M)
dimethoate, Cygon® (OP)	IPC, propham (M)
endosulfan, Thiodan® (CH)	malathion (OP)
fenthion, Baytex® (OP)	maneb (M)
heptachlor (CH)	methoxychlor (CH)
lindane (CH)	mirex (CH)
methomyl, Lannate® (C)	NAA (M)
naled, Dibrom® (OP)	oxythioquinox, Morestan® (M)
oxydemetonmethyl, Meta-Systox®-R (OP)	Perthane® (CH)
	piperonyl butoxide (M)
SMDC, Vapam® (M)	ronnel, Korlan® (OP)
strobane (CH)	rotenone (M)
trichlorfon, Dipterex®, Dylox® (OP)	simazine (M)
toxaphene (CH)	stirofos, Gardona® (OP)
	2,4,5-T (CH)
	tetradifon, Tedion® (CH)
	thiram (M)
	zineb (M)
	ziram (M)

[a] The estimates of hazards in this table are based primarily on the observed acute dermal and to a less extent oral toxicity of these compounds to experimental animals. Where it is available, use experience has also been considered. It should be noted that the classification into toxicity groups is both approximate and relative. These toxicity categories are not related to specific categories spelled out for label requirements.

[b] The chemical class to which the pesticide belongs is designated as follows: C, carbamate; CH, chlorinated hydrocarbon; F, formamidine; M, miscellaneous; N, nitro; and OP, organic phosphorus.

[c] The fumigant compounds acrylonitrile, D-D®, and Telone® have systemic toxicities that would indicate their placement under the "Less Dangerous" category. However, special note should be taken of the fact that the volatility of these compounds and their capacity to produce irritation of skin, eyes, and other tissues indicate that appropriate caution should be exercised in their use.

Source: H. R. Wolfe and W. F. Durham. 1966. "Safety in the Use of Pesticides." Proceedings, Second Eastern Washington Fertilizer and Pesticide Conference, Washington State University, Pullman, Washington, pp. 14–21. Also in J. B. Baily and J. E. Swift. 1968. *Pesticide Information and Safety Manual.* University of California Agricultural Extension Service, Berkeley, p. 18.

TABLE 10
Examples of toxicity classes.

Label classification	Oral LD$_{50}$	mg/kg	Dermal LD$_{50}$	mg/kg
		INSECTICIDES		
Highly toxic	aldicarb, Temik®	0.65–0.79	parathion	7–21
	TEPP	1.0	mevinphos, Phosdrin®	4.2–4.7
	monocrotophos, Azodrin®	21		
	phorate, Thimet®	1.1–2.3		
Moderately toxic	propoxur, Baygon®	95–104	methyl parathion	67
	DDT	113–118	dioxathion, Delnav®	63–235
			azinphosmethyl, Guthion®	220
Slightly toxic	malathion	1,000–1,375	toxaphene	780–1,075
	carbaryl, Sevin®	500–850	mexacarbate, Zectran®	1,500–2,500
			dicofol, Kelthane®	1,000–1,230
			malathion	>4,444[a]
			carbaryl, Sevin®	>4,000
		HERBICIDES		
Highly toxic	acrolein	46		
	sodium arsenite	10–50		
Moderately toxic	2,4-D	375	pentachlorophenol	150–350
	Paraquat	157	allyl alcohol	89
Slightly toxic	MSMA	700–1,800	endothall	750
	monuron	2,300–3,700	dichlobenil	500
			2.4-D acid	1,500
			MCPA	>1,000
		FUNGICIDES		
Highly toxic	cycloheximide, Actidione®	1.8–2.5		
	dinoseb	37–60		
Moderately toxic	binapacryl	58–63	pentachlorophenol	150–350
	tryphenyltin hydroxide	108	sodium pentachlorophenate	257
Slightly toxic	thiram	780	binapacryl	720–810
	anilazine	2,710	dinoseb	500
			maneb	>1,000
			zineb	>1,000
			triphenyltin hydroxide	5,000

[a] > means LD$_{50}$ is higher than figure shown.

FIELD REENTRY SAFETY INTERVALS

Because insecticides pose the greatest health hazard to the agricultural worker, with the organophosphate insecticides being the most important chemical group in this respect, field reentry safety intervals have been established.

The EPA now requires safety waiting intervals between application of certain insecticides and worker reentry into all treated fields, to prevent unnecessary exposure. Several states (for example, California) have adopted waiting intervals longer than those required by the EPA. The waiting intervals established by EPA are

48 HOURS
Ethyl parathion
Methyl parathion
Demeton (Systox®)
Monocrotophos (Azodrin®)
Carbofenothion (Trithion®)
Oxydemetonmethyl (MetaSystox-R)
Dicrotophos (Bidrin®)
Endrin

24 HOURS
Azinphosmethyl (Guthion®)
Phosalone (Zolone®)
EPN
Ethion

For all other insecticides, it is necessary only that workers wait until sprays have dried or dusts have settled before reentering treated fields. These worker safety intervals are not to be confused with the familiar harvest intervals—the minimum days from last treatment to harvest—indicated on the insecticide label.

If it is necessary for workers to enter fields earlier than the required waiting intervals, they must wear protective clothing. This consists of a long-sleeved shirt, long-legged trousers or coveralls, hat, shoes, and socks.

These waiting intervals should not impose any undue hardship on pest-management specialists and agricultural pest control advisors, because application of any one of these materials would preclude the necessity for field inspection within the required waiting intervals.

FIRST AID FOR PESTICIDE POISONING

Poisoning symptoms may appear almost immediately after exposure or may be delayed for several hours, depending on the chemical, dose, length of exposure, and the individual. These symptoms may include, but are not restricted to, headache, giddiness, nervousness, blurred vision, cramps, diarrhea, a feeling of general numbness, or abnormal size of eye pupils. In some instances, there is excessive sweating, tearing, or mouth secretions. Severe cases of poisoning may be followed by nausea and vomiting, fluid in the lungs, changes in heart rate, muscle weakness, breathing difficulty, confusion, convulsions, coma, or death. However, pesticide poisoning may mimic brain hemorrhage, heat stroke, heat exhaustion, hypoglycemia (low blood sugar), gastroenteritis (intestinal infection), pneumonia, asthma, or other severe respiratory infections.

Regardless of how trivial the exposure may seem, if poisoning is present or suspected, obtain medical advice at once. If a physician is not immediately available by phone, take the person directly to the emergency ward of the nearest hospital and take along the pesticide label and telephone number of the nearest poison control center.

First-aid treatment is extremely important, regardless of the time that may elapse before medical treatment is available. The first-aid treatment received during the first 2 to 3 minutes following a poisoning accident may very well spell the difference between life and death.

FIRST AID IN THE EVENT OF CHEMICAL POISONING

If You Are Alone with the Victim

> *First*—See that the victim is breathing; if not, give artificial respiration.
> *Second*—Decontaminate him immediately; i.e., wash him off thoroughly. Speed is essential!
> *Third*—Call your physician.
>
> *Note:* Do *not* substitute first aid for professional treatment. First aid is only to relieve the patient before medical help is reached.

If Another Person Is with You and the Victim

Speed is essential; one person should begin first-aid treatment, while the other calls a physician.

The physician will give you instructions. He will very likely tell you to get the victim to the emergency room of a hospital. The equipment needed for proper treatment is there. Only if this is impossible should the physician be called to the site of the accident.

General

1. Give mouth-to-mouth artificial respiration if breathing has stopped or is labored.

2. Stop exposure to the poison and if poison is on skin cleanse the person, including hair and fingernails. If swallowed, induce vomiting as directed (see section on swallowed poisons).

3. Save the pesticide container and material in it if any remains; get readable label or name of chemical(s) for the physician. If the poison is not known, save a sample of the vomitus.

Specific

POISON ON SKIN

1. Drench skin and clothing with water (shower, hose, faucet).
2. Remove clothing.
3. Cleanse skin and hair thoroughly with soap and water; rapidity in washing is most important in reducing extent of injury.
4. Dry and wrap in blanket.

POISON IN EYE

1. Hold eyelids open, wash eyes with gentle stream of clean running water immediately. Use copious amounts. Delay of a few seconds greatly increases extent of injury.
2. Continue washing for 15 minutes or more.
3. Do *not* use chemicals or drugs in wash water. They may increase the extent of injury.

INHALED POISONS (DUSTS, VAPORS, GASES)

1. If victim is in enclosed space, do not go in after him without air-supplied respirator.
2. Carry patient (do not let him walk) to fresh air immediately.
3. Open all doors and windows, if any.
4. Loosen all tight clothing.
5. Apply artificial respiration if breathing has stopped or is irregular.
6. Call a physician.
7. Prevent chilling (wrap patient in blankets but do not overheat him).
8. Keep patient as quiet as possible.
9. If patient is convulsing, watch his breathing and protect him from falling and striking his head on the floor or wall. Keep his chin up so his air passage will remain free for breathing.
10. Do not give alcohol in any form.

SWALLOWED POISONS

1. CALL A PHYSICIAN IMMEDIATELY.
2. *Do not induce vomiting if:*
   a. Patient is in a coma or unconscious.
   b. Patient is in convulsions.
   c. Patient has swallowed petroleum products (that is, kerosene, gasoline, lighter fluid).
   d. Patient has swallowed a corrosive poison (strong acid or alkaline products)—symptoms: severe pain, burning sensation in mouth and throat.

3. If the patient can swallow after ingesting a corrosive poision, give the following substances by mouth. A corrosive substance is any material that in contact with living tissue will cause destruction of tissue by chemical action, such as lye, acids, Lysol, etc.
   a. *For acids:* Milk, water, or milk of magnesia (1 tablespoon to 1 cup of water);
   b. *For alkali:* Milk or water; for patients 1 to 5 years old, 1 to 2 cups; for patients 5 years and older, up to 1 quart.
   c. *Universal:* Condensed canned milk, as much as the victim can consume.

4. IF POSSIBLE, INDUCE VOMITING WHEN NONCORROSIVE SUBSTANCE HAS BEEN SWALLOWED.
   a. Give milk or water (for patient 1 to 5 years old, 1 to 2 cups; for patients over 5 years, up to 1 quart).
   b. Induce vomiting by placing the blunt end of a spoon, not the handle, or your finger at the back of the patient's throat, or by giving 2 tablespoons of salt in a glass of warm water (an emetic).
   c. When retching and vomiting begin, place patient face down with head lowered, thus preventing vomitus from entering the lungs and causing further damage. Do not let him lie on his back.
   d. Do not waste excessive time in inducing vomiting if the hospital is a long distance away. It is better to spend the time getting the patient to the hospital where drugs can be administered to induce vomiting and/or stomach pumps are available.
   e. Clean vomitus from person. Collect some in case physician needs it for chemical tests.

CHEMICAL BURNS OF SKIN

1. Wash with large quantities of running water.
2. Remove contaminated clothing.
3. Immediately cover with loosely applied clean cloth (any kind will do), depending on the size of the area burned.
4. Avoid use of ointments, greases, powders, and other drugs in first-aid treatment of burns.
5. Treat shock by keeping patient flat, keeping him warm, and reassuring him until arrival of physician.

CHAPTER 20

Safety is a state of mind.

World War II production safety slogan

Safe Handling and Storage of Pesticides

The first rule of safety in using any pesticide is to read the label and follow the directions and precautions printed on it. Pesticides are safe to use, provided common-sense safety precautions are practiced and provided they are used according to the label instructions. This especially means keeping them away from children and illiterate or mentally incompetent persons.

"Safety is a state of mind" is an old rule used in industrial safety engineering. But pesticide safety is more than a state of mind. It must become a habit with those who handle, apply, and sell pesticides—and certainly with those who supervise those who do. You can control pests in your home and garden with safety if you use pesticides properly.

SELECTION OF PESTICIDES

Before buying a pesticide, check the label. Make sure it lists the name of the pest you want to control. If in doubt, consult your county agent or other authority. Select the pesticide that is recommended by competent authority and consider the effects it may have on nearby plants and animals. Make certain that the label on the container is intact and up-to-date; it should include directions and precautions. And finally, purchase only the quantity needed for the current season.

MIXING AND HANDLING PESTICIDES

If the pesticide is to be mixed before applying, carefully read the label directions and current official state recommendations of the Cooperative Extension Service. This information can be obtained from your local county agent. It's always a good idea to wear rubber gloves when mixing pesticides and to stand upwind of the mixing container. Handle the pesticides in a well-ventilated area. Avoid dusts and splashes when opening containers or pouring into the spray apparatus. Do not use or mix pesticides on windy days. Measure the quantity of pesticide required accurately, using the proper equipment. Overdosage is wasteful; it will not kill more pests, it may be injurious to plants, and it may leave an excess residue on fruits and vegetables. Do not mix pesticides in areas where there is a chance that spills or overflows could get into any water supply. Clean up spills immediately. Wash pesticides off skin promptly with plenty of soap and water and change clothes immediately if they become contaminated.

APPLYING PESTICIDES

If the label calls for it, wear the appropriate protective clothing and equipment. Make certain that equipment is calibrated correctly and is in satisfactory working condition. Apply only at the recommended rate, and, to minimize drift, apply only on a calm day. Do not contaminate feed, food, or water supplies. This includes pet food and water bowls. Avoid damage to beneficial and pollinating insects by not spraying during periods when such insects are actively working on flowering plants. Honey bees as a rule are inactive at dawn and dusk, which is a good time for outdoor applications. Keep pesticides out of mouth, eyes, and nose. Do not use your mouth to blow out clogged hoses or nozzles. Observe precisely the waiting periods specified on the label between pesticide application and harvest of fruit and vegetables. Clean all equipment used in mixing and applying pesticides according to recommendations. Do not use the same sprayer for insecticide and herbicide applications.

After handling pesticides, wash the sprayer, protective equipment, and hands thoroughly. And, if you should ever become ill after using pesticides and believe you have the symptoms of pesticide poisoning, call your physician and take the pesticide label with you. This situation is highly unlikely, but it's always good to know about in an emergency.

STORING PESTICIDES

Around the home, the rule of thumb is to lock up all pesticides. Lock the room, cabinet, or shed where they are stored, to discourage children. Do not store pesticides where food, feed, seed, or water can be contaminated. Store in a dry, well-ventilated place, away from sunlight, and at temperatures above freezing. If your operation is larger than a typical homeowner's, mark all entrances to your storage area with signs bearing this caution: "PESTICIDES STORED HERE—KEEP OUT."

Keep pesticides only in original containers, closed tightly and labeled. Examine pesticide containers occasionally for leaks and tears. Dispose of leaking and torn containers and clean up spilled or leaked material immediately. Eliminate all outdated materials, and date the containers when purchased. And because many pesticide spray formulations are flammable, take precautions against potential fire hazards.

DISPOSING OF EMPTY CONTAINERS AND UNUSED PESTICIDES

Empty containers are never completely empty, so do not reuse them for any purpose. Instead, break glass containers, rinse metal containers twice with water, punch holes in top and bottom and leave in your trash barrels for removal to the official landfill trash dump. Empty paper bags and cardboard boxes should be torn or smashed to make unusable, placed in a larger paper bag, rolled, and relegated to the trash barrel. In summary, do not leave anything tempting in the trash barrel or dump. Those youthful alley raiders may be your own children!

UNWANTED PESTICIDES

First, offer to give unwanted pesticides to a responsible person in need of the materials. If this is not practicable, bury dry pesticides at a depth of at least 18 inches in a safe disposal site. Pour liquid pesticides into a pit dug in sandy soil. Do not take unwanted pesticides to an incinerator and do not incinerate them yourself.

There surely must be thousands of "do's" and "don'ts," sets of rules for handling pesticides that have generated over the past two decades with the increased awareness of pesticide hazards. The preceding gleanings are from many sources and may be useful to the reader in his or her own home situation, on the farm, in preparation for talks on safety, for inspections of schools and public buildings for safe playing and working conditions, or just for reference.

THE PESTICIDE LABEL

The single most important tool to the layman in the safe use of pesticides is the label on the container. The Federal Environmental Pesticide Control Act (FEPCA), which is discussed in Chapter 21, on pesticide laws, contains three very important points concerning the pesticide label that I feel should be further emphasized. They pertain to reading the label, understanding the label directions, and following these instructions carefully.

Two of the first provisions of FEPCA are that (1) the use of any pesticide inconsistent with the label is prohibited, and (2) deliberate violations by growers, applicators, or dealers can result in heavy fines or imprisonment or both. The third provision is found in the general standards for certification of commercial applicators that in essence licenses them to use *restricted-use* pesticides, namely in the area of label and labeling comprehension. For certification, applicators are to be tested on (1) the general format and terminology of pesticide labels and labeling; (2) the understanding of instructions, warning, terms, symbols, and other information commonly appearing on pesticide labels; (3) classification of the product (general or restricted use); and (4) the necessity for use consistent with the label.

FIGURE 5
Environmental Protection Agency format for general-use pesticide label. (*Source: Pesticide Registration Guidelines, U.S. Environmental Protection Agency, 1975.*)

In Figures 5 and 6 are shown the format labels for *general-use* and *restricted-use* pesticides as required by EPA to appear on all containers beginning in 1978. These labels are keyed as follows:

1. Product name.
2. Company name and address.
3. Net contents.
4. EPA pesticide registration number.
5. EPA formulator manufacturer establishment number.

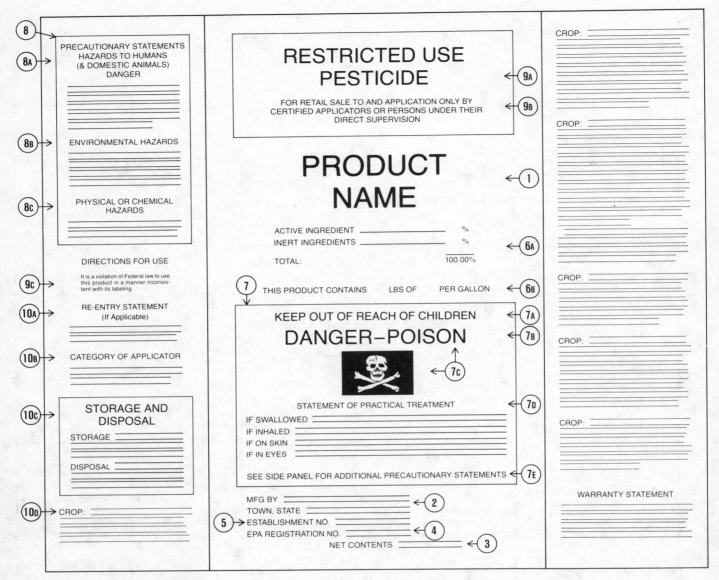

FIGURE 6
Environmental Protection Agency format for restricted-use pesticide label. Products bearing this kind of label are not available to the layperson. (*Source:* Pesticide Registration Guidelines, U.S. Environmental Protection Agency, 1975.)

6A. Ingredients statement.

6B. Pounds/gallon statement (if liquid).

7. Front-panel precautionary statements.

7A. Child hazard warning, "Keep Out of Reach of Children."

7B. Signal word—*DANGER, WARNING,* or *CAUTION.*

7C. Skull and crossbones and word *Poison* in red.

7D. Statement of practical treatment.

7E. Referral statement.

8. Side- or back-panel precautionary statements.
8A. Hazards to humans and domestic animals.
8B. Environmental hazards.
8C. Physical or chemical hazards.
9A. "Restricted Use Pesticide" block.
9B. Statement of pesticide classification.
9C. Misuse statement.
10A. Reentry statement.
10B. Category of applicator.
10C. "Storage and Disposal" block.
10D. Directions for use.

PESTICIDE EMERGENCIES

All insecticides can be used safely, provided common-sense safety is practiced and provided they are used according to the label instructions; this includes keeping them away from children and illiterate or mentally incompetent persons. Despite the most thorough precautions, accidents will occur. The following two sources of information are important in the event of any kind of serious pesticide accident.

The first and most important source of information is the CHEM-TREC telephone number. From this toll-free long-distance number can be obtained emergency information on all pesticide accidents, pesticide-poisoning cases, pesticide spills, and pesticide-spill cleanup teams. This telephone service is available 24 hours a day. The toll-free number is: CHEMTREC (800) 424-9300

The second source of information is only for human-poisoning cases: It is the nearest Poison Control Center. Look it up in the telephone directory under Poison Control Centers, or ask the telephone operator for assistance. Poison Control Centers are usually located in the larger hospitals of most cities and can provide emergency treatment information on all types of human poisoning, including pesticides. The telephone number of the nearest Poison Control Center should be kept as a ready reference by parents of small children or employees of persons who work with pesticides and other potentially hazardous materials.

Specific pesticide poisoning information can be obtained in writing or by telephone from: National Clearinghouse for Poison Control Centers, U.S. Department of Health, Education and Welfare, Food and Drug Administration, Bureau of Drugs, 5401 Westbard Avenue, Bethesda, Maryland 20016.

CHAPTER 21

There is hereby established the
Environmental Protection Agency.

"Presidential Documents,"
Federal Register, October 6, 1970

The Law and Pesticides

Federal laws have protected the user of pesticides, his pets and domestic animals, his neighbor, and the consumer of treated products for quite some time. Nothing is left unprotected.

The first federal law—the Federal Food, Drug, and Cosmetic Act of 1906, known as the Pure Food Law—required that food (fresh, canned, and frozen) shipped in interstate commerce be pure and wholesome. There was not a single word in the law that pertained to pesticides.

The Federal Insecticide Act of 1910, which covered only insecticides and fungicides, was signed into law by President Taft. The act was the first to control pesticides and was designed mainly to protect the farmer from substandard or fraudulent products, which were plentiful at the turn of the century. This was probably one of our earliest consumer protection laws.

The Pure Food Law of 1906 was amended in 1938 to include pesticides on foods, primarily the arsenicals, such as lead arsenate and Paris green. It also required the adding of color to white insecticides, including sodium fluoride and lead arsenate, to prevent their accidental use as flour or other look-alike cooking materials. This was the first federal effort toward protecting the consumer from pesticide-contaminated food, by providing tolerances for pesticide residues, namely arsenic and lead, in foods where these materials were necessary for the production of a food supply.

The Federal Insecticide, Fungicide, and Rodenticide Act (FIFRA) became law in 1947. It superseded the 1910 Federal Insecticide Act, extended the coverage to include herbicides and rodenticides, and required that any of these products must be registered with the U.S. Department of Agriculture before they could be marketed in interstate commerce. Basically, the law was one requiring good and useful labeling, making the product safe to use if label instructions were followed. The label was required to contain the manufacturer's name and address; name, brand, and trademark of the product; its net contents; an ingredient statement; an appropriate warning statement to prevent injury to humans, animals, plants, and useful invertebrates; and directions for use adequate to protect the user and the public.

The Miller Amendment to the Food, Drug and Cosmetic Act (1906, 1938) was passed in 1954. It provided that any raw agricultural commodity may be condemned as adulterated if it contains any pesticide chemical whose safety has not been formally cleared or that is present in excessive amounts (above tolerances). In essence, this amendment clearly set tolerances on all pesticides in food products; for example, 7.0 ppm DDT in lettuce, or 1.0 ppm ethyl parathion on string beans.

Two laws, the Federal Insecticide, Fungicide and Rodenticide Act (FIFRA) and the Miller Amendment to the Food, Drug, and Cosmetic Act, supplement each other and are interrelated by law in practical operation. Today they serve as the basic elements of protection for the applicator, the consumer of treated products, and the environment, as modified by the following amendments.

In 1958, the Food Additives Amendment to the Food, Drug and Cosmetic Act (1906, 1938, 1954) was passed. It extended the same philosophy to all types of food additives that had been applied to pesticide residues on raw agricultural commodities by the 1954 amendment. However, this also controlled pesticide residues in processed foods that had not previously fitted into the 1954 designation of raw agricultural commodities. Of greater importance, however, was the inclusion of the Delaney clause, which states that any chemical found to cause cancer (a carcinogen) in laboratory animals when fed at any dosage may not appear in foods consumed by humans. This has become the most controversial segment of the entire spectrum of federal laws applying to pesticides, mainly with regard to the dosage found to produce cancer in experimental animals.

The various statutes mentioned so far apply only to commodities shipped in interstate commerce. In 1959, FIFRA (1947) was amended to include nematicides, plant regulators, defoliants, and desiccants as economic poisons (pesticides). (Poisons and repellents used against amphibians, reptiles, birds, fish, mammals, and invertebrates have since been included as economic poisons.) Because FIFRA and the Food, Drug and Cosmetics Act are allied, these additional economic poisons were also controlled as they pertain to residues in raw agricultural commodities.

In 1964, FIFRA (1947, 1959) was again amended to require that all pesticide labels contain the federal registration number. It also required caution words, such as *Warning, Danger, Caution,* and *Keep Out of Reach of Children,* to be included on the front label of all poisonous pesticides. Manufacturers also had to remove safety claims from all labels.

The administration of FIFRA was the responsibility of the Pesticides Regulation Division of the U.S. Department of Agriculture until December 1970. At that time, the responsibility was transferred to the newly designated Environmental Protection Agency (EPA). Simultaneously, the authority to establish pesticide tolerances was transferred from the Food and Drug Administration (FDA) to EPA. The enforcement of tolerances remains the responsibility of the FDA.

In 1972, FIFRA (1947, 1959, 1964) was revised by the most important pesticide legislation of this century: the Federal Environmental Pesticide Control Act (FEPCA), sometimes referred to as the 1972

FIFRA Amendment. Some of the provisions of FEPCA are abstracted, as follows:

1. Use of any pesticide inconsistent with the label is prohibited.
2. Deliberate violations of FEPCA by growers, applicators, or dealers can result in heavy fines and/or imprisonment.
3. All pesticides will be classified into (a) general use or (b) restricted-use categories by October 1978.
4. Anyone applying restricted-use pesticides must be certified by the state in which he lives.
5. Pesticide manufacturing plants must be registered and inspected by EPA.
6. States may register pesticides on a limited basis when intended for special local needs.
7. All pesticide products must be registered by EPA, whether shipped in interstate or intrastate commerce.
8. For a product to be registered, the manufacturer is required to provide scientific evidence that the product, when used as directed, will (a) effectively control the pests listed on label, (b) not injure humans, crops, livestock, wildlife, or damage the total environment, and (c) not result in illegal residues in food or feed.

FEPCA established ten categories of certification for commercial applicators: (1) agricultural pest control (plant and animal); (2) forest pest control; (3) ornamental and turf pest control; (4) seed treatment; (5) aquatic pest control; (6) right-of-way pest control; (7) industrial, institutional, structural, and health-related pest control; (8) public health pest control; (9) regulatory pest control; and (10) demonstration and research pest control.

FEPCA also set general standards of knowledge for all categories of certified commercial applicators. Testing will be based, among other things, on the following areas of competency: (1) label and labeling comprehension, (2) safety, (3) environment, (4) pests, (5) pesticides, (6) equipment, (7) application techniques, and (8) laws and regulations. Much of this material is covered in this book.

These are only the most important aspects of FEPCA that you, the interested novice, need be acquainted with.

Beyond the federal laws providing rather strict control over the use of insecticides, each state usually has two or three similar laws controlling the application of pesticides and the sales and use of pesticides. These may or may not involve the licensing of aerial and ground applicators as one group and the structural applicator or pest control applicator as another.

PART VII

EPILOG

CHAPTER 22

Pesticides of the Future

This means that we have to find in the next 25 years, food for as many people again as we have been able to develop in the whole history of man.

Jean Mayer, 1975

What we are about to discuss is not speculative, for the pesticides of the future are here now, in their early stages of development. Let's examine some of the exciting activity that is grasping the imagination of research scientists in experimental field studies and in the laboratory.

INSECT CONTROL

Not all insect-control chemicals are insecticides. The discussions of the four following topics represent the fruition of thinking that began after the Civil War in the United States, with the introduction of arsenicals as stomach poisons for insects—the first true insecticides. This was the beginning of manipulating the environment of the insect to human advantage.

Insect Pheromones

It appears now that most insects communicate in some manner with chemicals. They release ultrasmall quantities of highly specific compounds that vaporize readily and are detected by insects of the same species. As we learn more about insect behavior, we will discover that all insects communicate through the release of these delicate molecules, known as *pheromones*.

Probably the most potent physiologically active molecules known today are insect pheromones. The word *pheromone* comes from the Greek *pherein*, "to carry," and *horman*, "to excite or stimulate." Pheromones are excreted to the outside of the insect body, where they cause specific reactions from other insects of the same species; they are also referred to in older literature as *social hormones*.

Pheromones can be classified into the following behavioral categories, based on the behavioral response of the receiving insect: sexual behavior, aggregation (including trail following), dispersion, oviposition, alarm, and specialized colonial behavior.

Of the different types of pheromones, it is the sex pheromones that offer the greatest potential to insect control. A good example of their recent use took place in the Texas high plains boll weevil-diapause control zone. Pheromone traps using live male boll weevils captured sufficient overwintered weevils to suppress the population until the late summer migration of large numbers of weevils overpowered the action of the traps.

The sex pheromones of pest moths have received the most detailed chemical study to date. For instance, after 30 years of study, the gypsy-moth sex pheromone was isolated, identified, and synthesized in the laboratory in 1960. Since then great quantities of disparlure, the synthetic female gypsy-moth sex pheromone, have been used in male-trapping programs for this forest pest.

A current list of the available synthetic sex pheromones is presented in Table 11.

We have no examples of complete insect control using synthetic pheromones, but many species are under investigation. The uses of synthetic pheromones include attracting mate-seeking insects to mechanical or sticky traps, to insecticide-treated areas, to poisoned baits, or to ultraviolet light traps. A method under study is designed to distribute gossyplure, the pink bollworm sex pheromone, in cotton fields in the spring to confuse emerging males and prevent mating. This principle could conceivably be applied to other insect species when synthetic pheromones become available in large quantities.

Despite the exuberance with which the potentials of sex pheromones have been praised, pheromones are most practically used in survey traps to provide information about population levels,

TABLE 11
Some of the commercially available synthetic insect sex pheromones.

Common name	Trade name	Species attracted
disparlure	Disparmone Pherocon GM	gypsy moth
grandlure	Grandamone Pherocon BW	boll weevil
gossyplure	Pherocon PBW	pink bollworm
hexalure[a]	Hexamone	pink bollworm
looplure	Cabblemone Pherocon CL	cabbage looper
muscalure	Muscamone	house fly
codlelure	Codlemone	codling moth
virelure	none	tobacco budworm
none	Z-11	red-banded leaf roller European corn borer oblique-banded leaf roller smartweed borer

[a] Obsolete.

to delineate infestations, to monitor control or eradication programs, and to warn of new pest introductions. It is likely that insect control systems of the future will rely heavily on the uses of pheromones, as a survey tool and to suppress early-emerging populations through trapping.

CHEMOSTERILANTS

Chemicals used to sterilize insects, thus preventing reproduction, are known as *chemosterilants*. More than 1,000 compounds that affect reproduction in insects have been described that come under the very broad classification of chemosterilants. The massive research effort to uncover these sterilizing chemicals is a direct spin-off of the success of programs for the eradication of the screwworm, a severe pest of beef cattle, by the release of males sterilized by gamma radiation. This concept involves the use of insects for the destruction of their own species through the induction of sterility in a large proportion of the males. It utilizes their mating behavior to reach female insects that would not be affected by the usual insecticidal control techniques.

The advantages of using chemicals instead of rearing, irradiating, and releasing massive numbers of males to induce sterility in a population are obvious. When it was shown that both males and females could be sterilized simultaneously, the potential seemed to be limitless. The development of chemosterilants is relatively new, and no definitive statements can be made about all types of compounds with sterilizing activity.

Chemosterilants are divided into four classes: alkylating agents, phosphorus amides, triazines, and antimetabolites.

The alkylating agents constitute the largest and most active class. They are moderately to highly reactive compounds with proteins and nucleic acids. They are sometimes referred to as *radiomimetics* (radiation-mimicking materials) in that their effects are similar to those of X rays or gamma rays. These agents replace hydrogen in fundamental genetic material with an alkyl group ($-CH_3$ or $-C_2H_5$, and so on), which results in an effect similar to irradiation. They are highly effective in producing mitotic disturbances or nucleotoxic conditions, particularly in tissues where cell division and multiplication take place at a high rate. This results in the production of multiple dominant lethal mutations or severely injured genetic material in the sperm or the egg. Although fully alive, zygotes (fertilized eggs), if formed, do not complete development into mature progeny.

The two most widely investigated types of alkylating chemosterilants are aziridines and alkanesulfonates. The chemical and physical properties of these compounds are quite variable, but their cytotoxic and mutagenic effects are closely related. Although the alkylating agents are relatively unstable and degrade rapidly, the possible contamination of large areas, even with small residues, makes their use as crop sprays or dusts hazardous and undesirable. Safe applications are possible, however, when these chemosterilants are used to sterilize reared or collected insects under controlled conditions and when personnel are adequately protected. Alkylating agents shown in Table 12 are apholate, tepa, metepa, thiotepa, tretamine, and busulfan.

The alkylating chemosterilants have been used in the field with moderate to good success for experimental housefly control around garbage and trash dumps. Busulfan fed to boll weevils that were later released in the field resulted in only moderate success. Numerous laboratory studies indicate great potential for the chemosterilant principle, but the obvious problem is the hazardous nature of the residues on food and feed crops. Again, we are dealing here with another form of chemical control, where modes of action differ from those of conventional insecticides.

MICROBIALS

This subject has been touched on in the chapter on insecticides. Microbials are disease organisms of insects. The term *microbial control* refers to the control of insect pest infestations using these disease organisms. In one sense, it is inappropriately placed in the "potential" section of this book. It more correctly belongs in a transitory section, indicating its current limited use as well as its vast potential for the future. Insect pathology is a relatively new field, even though the idea has germinated slowly over hundreds of years.

Microbial control as a potential tool in insect control programs owes allegiance to two currently used methods, chemical and biological controls. Some of the advantages and disadvantages of both methods apply also to microbial control. For example, some pathogens can be mass-produced (as are chemical insecticides in chemical control), applied in a conventional manner at certain dosage levels to kill an existing infestation, and dissipated in the environment. In such cases, the microbial agent is essentially a "living insecticide," and no prolonged or residual effects of the application are expected.

Because microbial agents are living organisms, many of the principles that apply to other biological control agents, such as parasites and predators, apply equally as well to pathogens; for example, they may be introduced into an environment to initiate a disease outbreak. But the main effects of the pathogens come from reproduction and spread of the disease organisms in the pest population. In other words, as with parasites and predators, they are self-perpetuating and regulatory in nature. They remain in the environment and become a permanent mortality factor in the pest population. Good examples of this are the milky (spore) disease bacteria *Bacillus popilliae* Dutky and *B. lentimorbus* Dutky, used to control the Japanese beetle.

The bacterium *Bacillus thuringiensis* Berliner is a pathogen used for insect control in a manner similar to the use of conventional insecticides. It is the only microbial agent now registered for use on food crops in the United States. Since it is short-lived, repeated applications are necessary.

Disease-causing bacteria, viruses, fungi, nematodes, and protozoa affect a wide range of insects, beneficials and pests alike. In nature,

TABLE 12
Common and chemical names, mammalian oral LD$_{50}$s, and chemical structures of the more common chemosterilants.

Common and chemical names	Oral LD$_{50}$ mg/kg	Chemical structure
Apholate 2,2,4,4,6,6-hexakis(1-aziridinyl)-2,2,4,4,6,6-hexahydro-1,3,5,2,4,6-triazatri-phosphorine	98	
Tepa tris(1-aziridinyl)phosphine oxide	37	
Metepa tris(2-methyl-1-aziridinyl)phos-phine oxide	136	
Thiotepa tris(1-aziridinyl)phosphine sulfide	9	
Tretamine 2,4,6,-tris(1-aziridinyl)s-triazine	1	
Busulfan 1,4-butanediol dimethane-sulfonate	18	$H_3C-S-O-C_4H_8-O-S-CH_3$
Hempa hexamethylphosphoric triamide	2,650 ♂ 3,360 ♀	
Thiohempa hexamethylphosphorothioic triamide	20	
Hemel hexamethylmelamine	350	

they play a large role in regulating insect pest numbers. In most agroecosystems, practically no year passes without almost complete decimation of cabbage loopers by a naturally occurring nuclear polyhedrosis virus. Similarly, a polyhedrosis virus of the alfalfa caterpillar is important in the natural control of this pest. Under favorable humidity and temperature conditions, pathogenic fungi play an important part in the natural control of a wide variety of insects. For example, several species of fungi are parasitic on the spotted alfalfa aphid.

Although many insect pests are subject to mortality from pathogenic agents occurring naturally in the environment, little reliance can be placed on them because of their unpredictable nature. Much research has been done in an effort to understand better the relationship of the three important components of a disease outbreak: the host insect, the pathogen, and the environment. As information is accumulated on these organisms and as a better understanding of their ecological requirements is gained, the importance of microbial control in insect control programs will surely increase to the point of their becoming a major tool in the total management scheme. For example, some pathogens exhibit high virulence against certain pests in the laboratory, but under field conditions relatively little effect is observed. The virus of the corn earworm is a case in point. The first attempts to utilize the nuclear polyhedrosis virus for field control of this pest were totally unsatisfactory, because ultraviolet radiation rendered the virus ineffective. Several formulations of this virus have been prepared, in an attempt to shield the virus particles from excessive radiation, and have increased its effectiveness in the field.

Although the use of B. thuringiensis for control of pest insects is beyond the "potential" stage, its full utilization lies ahead. The "potential" of the many other promising pathogens, particularly the viruses, also offers much hope for adding another important weapon to the insect control arsenal. This is true whether the pathogens more nearly resemble the conventional insecticides, as in chemical control, or the beneficial insects, as in biological control, in terms of their effects on pest populations.

INSECT GROWTH REGULATORS

Now we arrive at the third generation of insecticides, insect growth regulators (IGRs). This term encompasses a relatively new group of chemical compounds that alter growth and development in insects. Their effects have been observed on embryonic and larval and nymphal development, on metamorphosis, on reproduction in both males and females, on behavior, and on several forms of diapause. They include ecdysone (the molting hormone), juvenile hormone (JH), JH mimic, JH analog (JHA), and their broader synonyms, juvenoids and juvegens. More recently another growth effect, chitin inhibition, has been identified and is discussed later, with the EPA-registered IGRs.

First-generation insecticides are the stomach poisons, such as the arsenicals; the second generation includes the familiar organochlorine, organophosphate, carbamate, and formamidine insecticides. The third-generation insecticides, then, are the ultimate, required only in very minute quantities and apparently having no undesirable effects on humans and wildlife. They are, however, nonspecific, since they affect not only the target species, but most other arthropods as well. Consequently, when used with precision IGRs may play an important role in future insect control.

Several glands in insects are known to produce hormones, the principal functions of which are the control of reproductive processes, molting, and metamorphosis. Here we are interested only in the hormone ecdysone, which is responsible for molting, and JH, which inhibits or prevents metamorphosis.

When insects are treated with ecdysones, they usually die in all stages of growth, making ecdysones similar to second-generation insecticides. One attractive feature of ecdysones as potential tools is their widespread distribution in plants. More than 40 compounds have been isolated from higher plants, and these may play as yet unrecognized roles in insect-plant relationships. Some actually serve to inhibit the development of insects feeding thereon, thus protecting the host plant. These are referred to broadly as *antijuvenile hormones*, or, more accurately, *antiallatropins*.

Keen interest has been directed toward JHs. These are not, in the usual sense, toxic to insects. Instead of killing directly, they interfere in the normal mechanisms of development and cause the insects to die before reaching the adult stage.

Dramatic results have been obtained in the laboratory, the most promising effects being on mosquito larvae, caterpillars, and certain beetle larvae, although effects have been observed on practically all insect orders. Most insect species respond to treatment with JHs by producing extra larval, nymphal, or pupal forms that vary from giant, almost perfect, forms to intermediates of all sorts between the immatures and the adults. For the most part, the periods of greatest sensitivity for metamorphic inhibition are the last larval or nymphal stages and the pupa, in those having complete metamorphosis. One recognizable problem is the precision of timing applications to achieve maximum damaging effect on the upcoming life stage of a particular insect.

For practical purposes, IGRs could be used on crops to suppress damaging insect numbers. They would be applied with the purpose of preventing pupal development or adult emergence, thus keeping the insects in the immature stages, resulting eventually in their deaths.

To date, there are only two IGRs registered by the EPA. The first, methoprene (Altosid®), manufactured by the Zoecon Corporation, was registered early in 1975 as a mosquito growth regulator, for use against second through fourth larval stage floodwater mosquitoes at 3 to 4 ounces per acre, to prevent adult emergence. Larvae exposed to Altosid continue their development to the pupal stage, when they die. Altosid has no effect when applied to pupae or adult mosquitoes. Currently its use is limited to public health officials and other trained personnel of public mosquito-abatement programs.

The second IGR is diflubenzuron (Dimilin®), manufactured by Thompson-Hayward Chemical Company, registered in 1976 for the control of gypsy moth caterpillars in forests. It is not truly a growth regulator of the juvenoid class but rather another insecticide. Because of its mode of action, however, it is tentatively classed with the IGRs. Dimilin acts on the larval stages of most insects by inhibiting or blocking the synthesis of chitin, a vital and almost indestructible part of the hard outer covering of insects, the exoskeleton. Dimilin has shown great promise as a mosquito larvicide in experiments using as little as 0.5 gram per acre!

Can IGRs become successful pest control agents? Certainly they can in time. They will, however, have to meet the general criteria for other pest control agents; thus, they must be effective in reducing insect populations below economic damage levels, be competitive with second-generation insecticides in cost, and have no undesirable side effects. In summary, it appears that IGRs hold intriguing possibilities for future use in practical insect control. It should also be kept in mind that IGRs are insect-controlling chemicals and thus fall within the same legal confines as other insecticides. Their great distinction, however is that they have modes of action different from the traditional insecticides.

FUNGICIDES

In 1970 the EPA banned the use of alkyl mercurial fungicides for seed treatment. This was, in part, due to Sweden's ban, when it was established that the decline of seed-eating and predatory bird species in Sweden was caused by the use of alkyl mercurials for seed treatment. In these two moves, plant disease control lost many of its best chemical agents, and immediately it became necessary to find substitutes.

There are two approaches to chemical control of plant diseases: protecting plants from infection or curing plants after they become infected. Historically, the preventive route has been the only method of control until the recent appearances of the systemic fungicides, the oxathiins and benzimidazoles in 1966 and the pyrimidines in 1968. These will undoubtedly be landmarks in the history of fungus control.

The systemics protect susceptible foliage and flowers more efficiently than the protectants because of their ability to translocate through the cuticle and across leaves. And, because of the great concern over agricultural chemical pollution, the systemics offer a new path for specific placement of pesticides. If the total dosage and number of treatments needed for control can be reduced, excess chemical use could be avoided. Systemics may also replace certain more hazardous materials.

Unfortunately, the systemics developed to date are so specific that resistance is inevitable. This, of course, means that their overall life expectancies will be much shorter than the traditional heavy-metal and broad-spectrum protectants. However, now that the principles of systemic fungicide structures and modes of action have been elucidated, it should be relatively simple to bring additional materials to commercial use.

The needs in plant disease control are systemics and curatives for the downy mildew diseases, good systemic bactericides, and compounds that translocate to new foliage or roots following foliar treatment. As these are eventually provided, we should begin to lay aside those in current use that suffer from the usual inadequacies.

The future fungicides, then, accent the present systemics currently in use with more of similar basic chemical structures and their high order of specificity. This in no way will detract from use of protectants, since these are the "old reliables" and form the foundations for thorough and efficient disease control accepted as routine in today's cropping practices.

HERBICIDES

Herbicides, like all other pesticide groups, require new, more effective replacements for those that, for one reason or another, lose their utility. As new chemicals are introduced and groups of weeds are put under control, other weeds very soon, being freed of competition and being tolerant of the chemical, take over and become serious pests. With the advent of the chlorophenoxy compounds (2,4-D, etc.), there was a shift from broad-leaf weeds to grassy ones. Similar shifts have occurred when one chemical group of herbicides has been used continuously. In some instances, this problem has been solved by using a rotation of herbicide groups. In other situations, mixtures of herbicides are used to broaden the spectrum of weeds that can be controlled.

The continuing increase in herbicide use and dollar investment indicates increased profits for agriculture and shows that chemical weed control, in addition to alleviating the tremendous burden of manual weeding, has increased the net income of growers around the world. Despite the advances of herbicide technology, improvements can be expected to continue for decades as herbicides become available and used in the Third World countries.

Underlying this growing usage of herbicides in agriculture lies a massive research effort, involved in synthesis, testing, development, and production of new herbicides. Techniques from practically every aspect of biology have been included in these activities. Biochemistry and plant physiology laboratories in universities and federal research facilities, as well as those of industry, have carried out research on the absorption, translocation, and mode of action of herbicides. Studies on the morphological effects of herbicides have been made. Soil science, microbiology, and pesticide residue laboratories have been involved in studies of the fate of herbicides, including absorption, conjugation, chemical alteration, and photolytical and biological degradation. Much of these studies represents the normal effort required to understand the functions of herbicides and secure their registrations under the ever-increasing demands of the EPA.

Herbicides that have produced residual problems, those that resist extensive degradation photolytically or biologically and remain in soils and plant products as intermediate breakdown products, are being phased out by EPA or closely scrutinized with that in mind. The history of the persistent insecticides is a good picture of the future for herbicides.

The herbicide of the future will undoubtedly be developed rather heterogeneously from the many chemical groups already in existence, as are those introduced within the last 3 years. Structures for 14 of the most recent, belonging to seven chemical classifications, are shown as follows.

Nitroanilines

PROSULFALIN (Sward®)

N-([4-(dipropylamino)-3,5-dinitrophenyl]-
sulfonyl)-S,S-dimethylsulfilimine

ETHALFLURALIN (Somilan®)

N-ethyl-N-(2-methyl-2-propenyl)-2,6 dinitro-
4-(trifluoromethyl)benzenamine

Substituted Amides

CGA-24705

2-chloro-N-(2-ethyl-6-methylphenyl)-N-
(2-methoxy-1-methylethyl)acetamide

VEL 5052

2-chloro-N-(2,6-dimethyl phenyl)-N-
(1,3-dioxolan-2-yl)methyl acetamide

Heterocyclic Nitrogens

DOWCO-290
(resembles Picloram)

3,6-dichloropicolinic acid

TRICLOPYR, DOWCO 233
(resembles Picloram)

[(3,5,6-trichloro-2-pyridyl)-oxy] acetic acid

Pyridiliums

CYPERQUAT, GCP-1634

1-methyl-4-phenylpyridinium chloride

Substituted Ureas

BAY-MET 1486

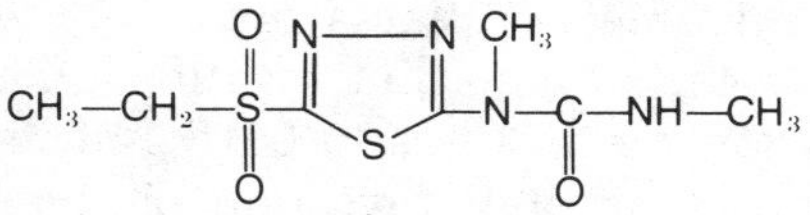

N-[5-ethylsulfonyl)-1,3,4-thiadiazol-2-yl]-
N,N'-dimethylurea

CISANILIDE (Rowtate®)
(resembles Monuron)

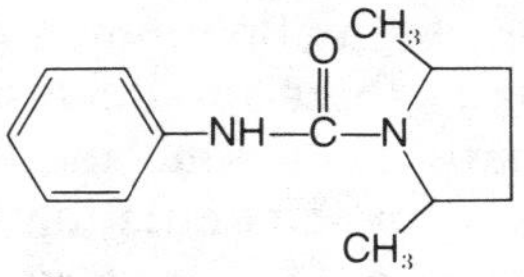

cis-2,5-dimethyl-1-pyrrolidinecarboxanilide

Phenol Derivatives

RH-2512

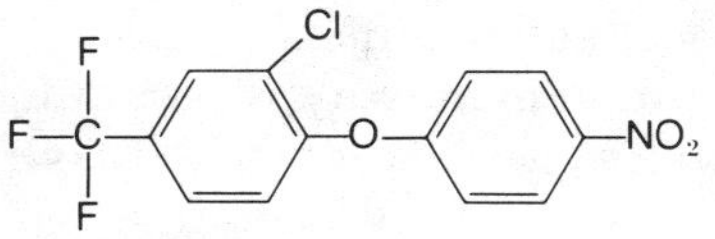

2-chloro-1-(4-nitrophenoxy)-4-
trifluoromethyl benzene

RH-2915

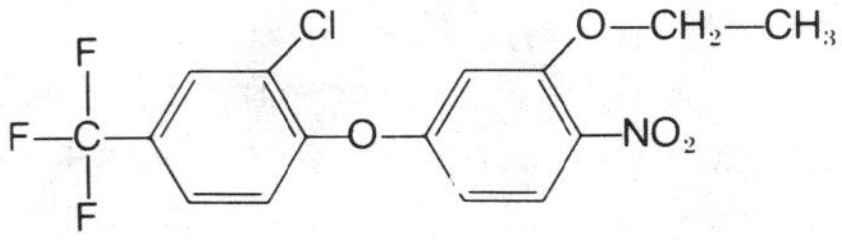

2-chloro-1-(3-ethoxy-4-nitrophenoxy)-4-
trifluoromethyl benzene

Organophosphates

BAY NTN 6867

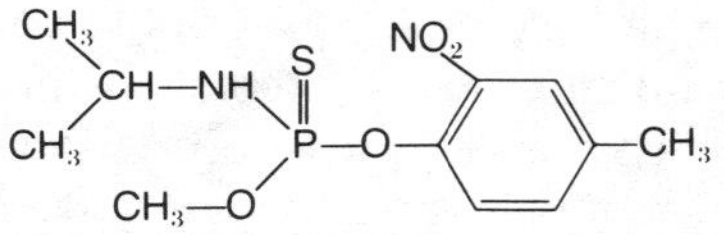

O-methyl O-(4-methyl-2-nitrophenyl)-
(1-methylethyl) phosphoramidothioate

ALTERNATIVES TO PESTICIDES:
WHERE DO WE GO FROM HERE?

It requires no wisdom to predict that we will continue to utilize pesticides heavily in the future, at least throughout our generation. Pesticides are essential and will remain our first line of defense against pests when damage levels reach economic thresholds. However, complete dependence on chemical control while ignoring all other methods of pest control has serious consequences, some of which are familiar.

In the mid-1960s, we heard words like *persistence* and *biomagnification* in relation to the chlorinated insecticides. Persistence is the quality of a compound to retain its chemical identity and biological effectiveness for long periods of time. This is considered highly desirable for continued pest control, but it also causes some environmental problems. A case in point, DDT.

DDT was probably the most useful and economical insecticide available in our history. Consequently, it was used not only in great quantity, but also in excessive quantity. Its persistence and fat solubility permitted it to accumulate in animals or plants and enter the food chain of both humans and wildlife. Humans and wildlife at the tops of these food chains received large exposures to DDT and its equally stable metabolite DDE simply through ingestion of their foods, an accumulation termed *biomagnification*.

Another familiar term is *resistance*. Most staphylococcus or staph infections are resistant to penicillin; flies have developed resistance to DDT; cockroaches have developed resistance to chlordane; some strains of gonorrhea cannot be controlled with any but the very latest antibiotics; some recently developed fungicides with specific modes of toxic action fail to control certain plant diseases; and certain weed species in given geographical areas have become very difficult to control with herbicides that were totally effective a decade ago.

Insect resistance to insecticides is a problem worthy of genuine concern. In 1944, only 44 insect species were known to have developed resistance to the available insecticides. Remember that the new synthetic insecticides were not yet on the scene. Today's estimates place this number greater than 250, half of which are of agricultural significance.

The last consequence familiar to the reader is the effects on non-target organisms. Insecticides are normally applied to an agricultural crop for only a few pest insects. In most cases, these few key species require this kind of artificial control measure to prevent economic losses to the crop. The history of insecticide usage, however, illustrates the fact that additional problems are created, either by the rapid resurgence of the treated pest population or by raising minor pests to the role of secondary or major pest status.

Insecticide applications not only reduce the pest population but populations of natural enemies of the pest as well, with a resultant increase in pest populations. Here is a good example. In 1946 and 1947, many thousands of acres of citrus trees in California developed damaging populations of the cottony-cushion scale following DDT applications for other pests. This was a result of the elimination of the vedalia beetle, a predator that had been the main factor controlling this scale since introduction of the beetle into California in 1888. Since then it has been demonstrated experimentally that elimination of vedalia beetle was the sole cause of the increase of the cottony-cushion scale.

There are many more examples of unintentional damage to non-target species; the drift of herbicides and insecticides onto sensitive crops or those intended for animal feeds; the reduction of beneficial soil microflora by application of fungicides or herbicides to the above-ground portions of the crop; the killing of pets and wildlife when baiting for rodent or predatory pests; the contamination of root crops from last year's insecticide or herbicide application; sickness and death of livestock feeding in pastures on which highly toxic materials drifted following application to adjacent crops; and, finally, the accidental poisoning of persons, including children, from improperly stored or secured pesticides.

The list goes on and on. The point is that pesticides have their consequences and the total dependence on pesticides results in increased human-made pest problems, environmental damage and potential hazard to humans themselves.

So what's the answer? The answer to long-range, intelligent pest control is the management, the manipulation of pests, using not just one, but all of the pest control methods available. This combination of all methods into one thoughtful program is referred to as *integrated pest management*, the practical manipulation of pests using any or all control methods in a sound ecological manner. This strategy brings together into a workable combination the best parts of all control methods that apply to a given problem created by the activities of pests.

Generally, control methods available today fall into the following categories, though all may not apply to every pest form:

1. *Chemical control*. The control method emphasized throughout this book, the use of pesticides.

2. *Biological control (biocontrol)*. The reduction of pest numbers by predators, parasites, or pathogens.

3. *Cultural control*. The use of farming or cultural practices associated with the crop production to make the environment less favorable for survival, growth, or reproduction of pest species.

4. *Host-plant resistance*. The resistance of plants to attack by insects, disease organisms, nematodes, or birds.

5. *Physical and mechanical control*. Direct or indirect measures that kill the pest, disrupt its physiology other than by chemical means, exclude it from an area, or adversely alter the pest's environment.

6. *Regulatory Control*. Preventing the entry and establishment of undesirable plant and animal pests in a country or area and eradicating, containing, or suppressing pests already established in limited areas (quarantines).

Most methods of pest control fall within these classifications. Theoretically, all crops and their pests lend themselves readily to integrated pest management. In reality, this is not yet true. The greatest advances have been made in insect pest management, followed by the management of plant pathogens and weeds. We have a long way to progress in this area.

In summary, our pesticide arsenal is not keeping up with our pest problems, because of resistance, persistence, hazards, and environmental complications. Whatever the cause of the declining availability of these precious chemical tools, we need to expand our horizons into integrated pest management to preserve their period of usefulness by using them when and only when they are needed. Our alternatives to this long-range plan are less than encouraging.

And, in closing, hear two more words of exhortation regarding chemicals and our environment: environmental respect. Regardless of your age, you have seen or read some of the undesirable aftereffects that chemicals in excess—not just pesticides, any chemicals—can have on our precarious environment: the effects of food-chain-incorporated DDE, the metabolite of DDT most frequently found in the environment, and polychlorinated biphenyls on eggshell thickness in birds of prey; marine oil spills with resulting fish kills and beach contamination; the reduction of ozone in the upper atmosphere by human-made fluorohydrocarbons; or contamination of rivers and streams by runoff of nitrogenous fertilizers and effluents from food-processing plants and the resulting reduction of aquatic life. These are but a few of the more bizarre, headline-rating episodes. We must, as a world culture, develop a greater respect for our delicate environment if the earth is to support the 7 billion residents anticipated by A.D. 2000.

APPENDIXES

Common, trade, and chemical names of insecticides and acaricides and their oral and dermal LD$_{50}$s to rats.

Common name, class[a], and trade name	Chemical name	Oral LD$_{50}$	Dermal LD$_{50}$
Abate® (*see* temephos)			
acephate (I) Orthene®	*O,S*-dimethyl acetylphosphoramidothioate	866	2,000
Actellic® (*see* pirimiphos-methyl)			
Agritox® (see *trichloronat*)			
Akton® (I) Shell SD-9098	*O*-[2-chloro-1-(2,5-dichlorophenyl)vinyl]*O,O*-diethyl phosphorothioate	146	177
aldicarb (I,M) Temik®	2-methyl-2-(methylthio)propionaldehyde *O*-(methylcarbamoyl)oxime	1	5
aldrin (I) Octalene®	1,2,3,4,10,10-hexachloro-1,4,4a,5,8,8a-hexahydro-1,4-*endo-exo*-5,8-dimethanonaphthalene	39	65
Alfacron® (*see* iodofenphos)			
allethrin (I)	*cis, trans*-(±)-2,2-dimethyl-3-(2-methylpropenyl)cyclopropane-carboxylic acid ester with (±)-2-allyl-4-hydroxy-3-methyl-2-cyclopenten-1-one	680	11,200
d-trans-allethrin (I) Bioallethrin	*trans*-(+)-2,2-dimethyl-3-(2-methylpropenyl)cyclopropanecar-boxylic acid ester with (±)-2-allyl-4-hydroxy-3-methyl-2-cyclopenten-1-one	425	4,000
Allied Chemical ACD-6506 (I)	dimethyl *p*-(methylthio)phenyl phosphate	7	46
Altosid® (*see* methoprene)			
Ambush® (*see* FMC 33297)			
amidithion (I,M) Thiocron®	*S*-[[2-methoxyethylcarbamoyl]methyl]*O,O*-dimethyl phosphorodithioate	600	1,600
aminocarb (I) Matacil®	4-(dimethylamino)-*m*-tolyl methyl carbamate	30	275
amitraz (M) Tactic®, Baam®	*N,N*′[(methylimino)dimethylidyne]bis[2,4-xylidine]	600	1,600
apholate (C)	2,2,4,4,6,6-hexakis(1-aziridinyl)-2,2,4,4,6,6-hexahydro-1,3,5,2,4,6-triazatriphosphorine	98	50
Aramite® (M)	2-(*p-tert*-butylphenoxy)-1-methylethyl-2-chloroethyl sulfite	3,900	—[b]
azinphosethyl (I) Ethyl Guthion®	*O,O*-diethyl phosphorodithioate *S*-ester with 3-(mercaptomethyl)-1,2,3-benzotriazine-4(3*H*)-one	7	80
azinphosmethyl (I) Guthion®	*O,O*-dimethyl phosphorodithioate *S*-ester with 3-(mercaptomethyl)-1,2,3-benzotriazine-4(3*H*)-one	13	220
Azodrin® (*see* monocrotophos)			
Baam® (*see* amitraz)			

(*continued*)

Common name, class,[a] and trade name	Chemical name	Oral LD$_{50}$	Dermal LD$_{50}$
Bacillus thuringiensis Biotrol® K, Dipel®, Thuracide®	Microbial insecticide for caterpillars	nontoxic	
Baygon® (*see* propoxur)			
BAY-NTN 9306 Bolstar®	O-ethyl O-[4(methylthio)phenyl]-S-propyl phosphorodithioate	227	>500
Bayrusil® (*see* quinalphos)			
Baytex® (*see* fenthion)			
bendiocarb (I) Ficam®	2,2-dimethyl-1,3-benzodioxol-4-yl-methylcarbamate	143	—
benzene hexachloride (I) BHC, HCH, 666	1,2,3,4,5,6-hexachlorocyclohexane	125	>4,000
Bidrin® (*see* dicrotophos)			
binapacryl (M) Morocide®	2-*sec*-butyl-4,6-dinitrophenyl 3-methyl-2-butenoate	136	1,010
Bioallethrin (*see* d-*trans*-allethrin)			
Biotrol® (*see* Heliothis virus)			
Bolstar® (*see* BAY-NTN 9306)			
bromophos (I) Nexion®	O-(4-bromo-2,5-dichlorophenyl) O,O-dimethyl phosphorothioate	3,750	2,188
bromophos-ethyl (I,M) Filariol®	O-(4-bromo-2,5-dichlorophenyl) O,O-diethyl phosphorothioate	52	1,366
bromopropylate (M) Acarol®	isopropyl 4,4'-dibromobenzilate	5,000	10,200
bufencarb (I) Bux®	3-(1-methylbutyl)phenyl methylcarbamate + 3-(1-ethylpropyl)phenyl methylcarbamate (3:1)	87	400
busulfan (C) Myleran®	1,4-butanediol dimethanesulfonate	18	—
butonate (I)	dimethyl (2,2,2-trichloro-1-hydroxyethyl)phosphonate butyrate	1,100	7,000
Bux® (*see* bufencarb)			
carbanolate (I,M) Banol®	6-chloro-3,4-xylyl methylcarbamate	31	—
carbaryl (I,M) Sevin®	1-naphthyl methylcarbamate	307	2,000
carbofuran (I) Furadan®	2,3-dihydro-2,2-dimethyl-7-benzofuranyl methylcarbamate	8	10,200
carbophenothion (I,M) Trithion®	S-[[(p-chlorophenyl)thio]methyl]O,O-diethyl phosphorodithioate	6	22
CGA-15324 Curacron®	O-(4-bromo-2-chlorophenyl) O-ethyl S-propyl phosphorothioate	400	1,610
chlorbenside (I,M) Mitox®	p-chlorobenzyl p-chlorophenyl sulfide	2,000	—
chlordane (I) Octachlor®	1,2,4,5,6,7,8,8-octachloro-3a,4,7,7a-tetrahydro-4,7-methanoindan	283	580
chlordimeform (I,M) Fundal®, Galecron®	N'-(4-chloro-o-tolyl)-N,N-dimethylformamidine	170	225
chlordimeform hydrochloride (I,M) Fundal® S.P.	N'-(4-chloro-o-tolyl)-N,N-dimethyl formamidine monohydrochloride	225	4,000
chlorfenethol (M) Dimite®	4,4'-dichloro-α-methylbenzhydrol	926	—

(continued)

Common name, class,[a] and trade name	Chemical name	Oral LD_{50}	Dermal LD_{50}
chlorfenvinphos (I,M) Supona®	2-chloro-1-(2,4-dichlorophenyl) vinyl diethyl phosphate	12	31
chlorobenzilate (M) Acaraben®	ethyl 4,4'-dichlorobenzilate	700	10,200
chloropicrin (F) Picfume®	trichloronitromethane	lachrymatory	
chloropropylate (M) Acaralate®	isopropyl 4,4'-dichlorobenzilate	5,000	150
chlorphoxim (I) Bay 78182	(o-chlorophenyl)glyoxylonitrile oxime O,O-diethyl phosphorothioate	—	—
chlorpyrifos (I) Dursban®, Lorsban®	O,O-diethyl O-(3,5,6-trichloro-2-pyridyl) phosphorothioate	97	2,000
chlorpyrifos-methyl (I) Dowco 214	O,O-dimethyl O-(3,5,6-trichloro-2-pyridyl) phosphorothioate	941	2,000
chlorthion (I) phosnichlor	O-(3-chloro-4-nitrophenyl O,O-dimethyl) phosphorothioate	880	—
Cidial® (see phenthoate)			
Ciodrin® (see crotoxyphos)			
Co-ral® (see coumaphos)			
coumaphos (I) Co-Ral®	O,O-diethyl O-(3-chloro-4-methyl-2-oxo-2H-1-benzopyran-7-yl)phosphorothioate	13	860
Counter® (see terbufos)			
crotoxyphos (I) Ciodrin®	α-methylbenzyl(E)-3-hydroxycrotonate dimethyl phosphate	125	385
crufomate (I) Ruelene®	4-tert-butyl-2-chlorophenyl methyl methylphosphoramidate	660	2,000
Curacron® (see CGA-15324)			
cyanthoate (M,I) Tartan®	O,O-diethyl phosphorothioate S-ester with N-(1-cyano-1-methylethyl-2-mercaptoacetamide	2	105
cyhexatin (M) Plictran®	tricyclohexylhydroxystannane	180	2,000
Cyolane® (see terbufos)			
Cytrolane® (see mephosfolan)			
Dasanit® (see fensulfothion)			
DBCP (F) Fumazone®, Nemagon®	1,2-dibromo-3-chloropropane	170–300	1,420
D-D® (see dichloropropane)			
DDT (I)	1,1,1-trichloro-2,2-bis(p-chlorophenyl)ethane	87	1,931
DDVP (see dichlorvos)			
Delnav® (see dioxathion)			
demeton (I) Systox®	mixture of O,O-diethyl-S(and O)-[2-(ethylthio)ethyl] phosphorothioates	2	8
demeton-methyl (see methyl demeton)			
dialifor (I,M) Torak®	O,O-diethyl phosphorodithioate S-ester with N-(2-chloro-1-mercaptoethyl) phthalimide	5	145
diazinon (I) Spectracide®	O,O-diethyl O-(2-isopropyl-6-methyl-4-pyrimidinyl) phosphorothioate	66	379

(continued)

Common name, class,[a] and trade name	Chemical name	Oral LD_{50}	Dermal LD_{50}
Dibrom® (*see* naled)			
dibromo-chloropropane (F) Nemagon®	1,2-dibromo-3-chloropropane	170–300	1,420
dicapthon (I)	*O*-(2-chloro-4-nitrophenyl) *O,O*-dimethyl phosphorothioate	475	>2,000
dichlofenthion (I) Nemacide®	*O*-2,4-dichlorophenyl *O,O*-diethyl phosphorothioate	270	6,000
o-dichlorobenzene (I)	*o*-dichlorobenzene	500	—
p-dichlorobenzene (I) PDB	*p*-dichlorobenzene	500	2,000
dichloroethyl ether (F) Chlorex®	bis(2-chloroethyl)ether	lachrymatory	
dichloropropane-dichloropropene (F) D-D	dichloropropane-dichloropropene mixture	140	2,100
1,3-dichloropropene (F) Telone®	1,3-dichloropropene	250	—
dichlorvos (I) DDVP, Vapona®	2,2-dichlorovinyl dimethyl phosphate	25	59
dicofol (M) Kelthane®	4,4′-dichloro-α-(trichloromethyl) benzhydrol	575	4,000
dicrotophos (I,M) Bidrin®	dimethyl phosphate ester with (*E*)-3-hydroxy-*N,N*-dimethylcrotonamide	22	225
dieldrin (I)	1,2,3,4,10,10-hexachloro-6,7-epoxy 1,4,4a,5,6,7,8,8a-octahydro-1,4-*endo-exo*-5,8-dimethanophthalene	40	65
diflubenzuron (GR) Dimilin®	*N*-(4-chlorophenyl-*N*′-2,6-difluorobenzoyl) urea	10,000	—
dimetan (I)	5,5-dimethyl-3-oxo-1-cyclohexen-1-yl-dimethylcarbamate	150	—
dimethoate (I) Cygon®	*O,O*-dimethyl *S*-(*N*-methylcarbamoylmethyl) phosphorodithioate	250	150
dimethrin (I)	2,4-dimethylbenzyl 2,2-dimethyl-3-(2-methylpropenyl)cyclo-propanecarboxylate	40,000	15,000
dimetilan (I)	1-(dimethylcarbamoyl)-5-methyl-3-pyrazolyl dimethylcarbamate	25	600
Dimilin® (*see* diflubenzuron)			
Dimite® (*see* chlorfenethol)			
dinitrobutylphenol (*see* dinoseb)			
dinitrocresol (I,M) DNOC	4,6-dinitro-*o*-cresol	26	200
dinitrocyclohexylphenol DNOCHP (I)	2-cyclohexyl-4-6-dinitrophenol	65	—
dinocap (M) Karathane®	2-(1-methylheptyl)-4,6-dinitrophenyl crotonate	980	4,700
dinoseb (I) DNOSBP	2-*sec*-butyl-4,6-dinitrophenol	37	80
dioxacarb (I) Famid®, Elocron®	*o*-1,3-dioxolan-2-ylphenyl methyl carbamate	125	1,950
dioxathion (I,M) Delnav®	2,3-*p*-dioxanedithiol *S,S*-bis(*O,O*-diethyl phosphorodithioate)	19	53
Dipel® (see *Bacillus thuringiensis*)			

(continued)

Common name, class,[a] and trade name	Chemical name	Oral LD$_{50}$	Dermal LD$_{50}$
Dipterex (see trichlorfon)			
disulfoton (I) Di-Syston®	O,O-diethyl S-[2-(ethylthio)ethyl] phosphorodithioate	2	20
Di-syston® (see disulfoton)			
DNC (see dinitrocresol)			
DNOC (see dinitrocresol)			
Dursban® (see chlorpyrifos)			
Dylox® (see trichlorfon)			
Dyfonate® (see fonofos)			
endosulfan (I) Thiodan®	6,7,8,9,10,10-hexachloro-1,5,5a,6,9,9a-hexahydro-6,9-methano-2,4,3-benzodioxathiepin 3-oxide	18	74
endothion (I) Exothion®	O-O-dimethyl phosphorothioate S-ester with 2-mercapto-methyl)-5-methoxy-4H-pyran-4-one	30	—
endrin (I)	1,2,3,4,10,10-hexachloro-6,7-epoxy-1,4,4a,5,6,7,8,8a-octahydro-1,4-endo-endo-5,8-dimethanonaphthalene	3	12
Entex® (see fenthion)			
EPN (I,M)	O-ethyl O-(p-nitrophenyl) phenylphosphonothioate	7	22
ethion (M,I) Nialate®	O,O,O'O'-tetraethyl S,S'-methylene bis(phosphorodithioate)	27	62
ethoprop (I) Mocap®	O-ethyl S,S-dipropyl phosphorodithioate	61	26
ethylene dibromide (I,F) EDB	1,2-dibromoethane		
ethylene dichloride (I,F) EDC	1,2-dichloroethane		
ethylene oxide (I,F)	ethylene oxide		
Ethyl Guthion® (see azinphosethyl)			
famphur (I) Warbex®	O,O-dimethyl O-[p-(dimethylsulfamoyl)phenyl] phosphorothioate	35	1,460
fenazaflor (M) Lovozal®	phenyl 5,6-dichloro-2-(trifluoromethyl)-1-benzimidazolecarboxylate	238	4,000
fenchlorphos (see ronnel)			
fenitrothion (I) Sumithion®	O,O-dimethyl O-(4-nitro-m-tolyl) phosphorothioate	250	200
fenson (M)	p-chlorophenyl benzenesulfonate	1,560	2,000
fensulfothion (I,M) Dasanit®	O,O-diethyl O-[p-(methylsulfinyl)phenyl]phosphorothioate	2	3
fenthion (I,M) Baytex®, Entex®	O,O-dimethyl O-[4-(methylthio)-m-tolyl] phosphorothioate	255	330
Ficam® (see bendiocarb)			
FMC 33297 (I) Ambush®, Pounce®	m-phenoxybenzyl cis,trans-(±)-3-(2,2-dichlorovinyl)-2,2-dimethylcyclopropanecarboxylate	2,000	—
Folimat® (see omethate)			
fonofos (I) Dyfonate®	O-ethyl S-phenyl ethyl phosphonodithioate	8	147

(continued)

Common name, class,[a] and trade name	Chemical name	Oral LD$_{50}$	Dermal LD$_{50}$
formetanate hydrochloride (I,M) Carzol® S.P.	m-[[(dimethylamino)methylene]amino]phenyl methylcarbamate monohydrochloride	15	5,600
formothion (I,M) Anthio®	O,O-dimethyl phosphorodithioate S-ester with N-formyl-2-mercapto-N-methylacetamide	365	1,680
fospirate (I,M) Dowco 217	dimethyl 3,5,6-trichloro-2-pyridyl phosphate	869	100
Fumazone® (see DBCP)			
Fundal® (see chlordimeform)			
Furadan® (see carbofuran)			
Galecron® (see chlordimeform)			
Genite® 923 (M) Genite®	2,4-dichlorophenyl benzenesulfonate	980	940
glyodin (M) Glyoxide®	2-heptadecyl-2-imidazoline acetate	6,800	—
Guthion® (see azinphosmethyl)			
Heliothis polyhedrosis virus (I) Biotrol®, Viron-H® VHZ	viral insecticide specific for Heliothis	nontoxic	
hemel (C)	hexamethylmelamine	350	—
hempa (C) HMPA	hexamethyl phosphoric triamide	<2,650	1,500
heptachlor (I)	1,4,5,6,7,8,8-heptachloro-3a,4,7,7a,tetrahydro-4,7-methanoindene	40	119
Hostathion® (see triazophos)			
hydrocyanic acid (F) hydrogen cyanide prussic acid	hydrocyanic acid	<0.5	—
Imidan® (see phosmet)			
iodofenphos (I) Alfacron®	O-(2,5-dichloro-4-iodophenyl)O,O-dimethyl phosphorothioate	2,000	1,800
isobenzan (I) Telodrin®	1,3,4,5,6,7,8,8-octachloro-1,3,3a,4,7,7a-hexahydro-4,7-methanoisobenzofuran	8	5
isodrin (I)	1,2,3,4,10,10-hexachloro-1,4,4a,5,8,8a-hexahydro-1,4-endo-endo-5,8-dimethanonaphthalene	24	85
Kepone® (I) chlordecone	decachlorooctahydro-1,3,4-methano-2H-cyclobuta[cd]pentalen-2-one	95	345
leptophos (I) Phosvel®	O-(4-bromo-2,5-dichlorophenyl)O-methyl phenylphosphonothioate	43	10,000
lindane (I) gamma-BHC	1,2,3,4,5,6-hexachlorocyclohexane, gamma isomer of not less than 99% purity	76	500
malathion (I,M) Cythion®	diethyl mercaptosuccinate S-ester with O,O-dimethyl phosphorodithioate	885	4,000
mecarbam (I) Murfotox®	S-[[(ethoxycarbonyl)methylcarbamoyl]methyl]O,O-diethyl phosphorodithioate	15	380
menazon (I,M) Sayfox®	S-[(4,6-diamino-S-triazin-2-yl)methyl]O,O-dimethyl phosphorodithioate	1,200	500
mephosfolan (I) Cytrolane®	cyclic propylene p,p-diethyl phosphonodithio imidocarbonate	9	100
Mesurol® (see methiocarb)			
metepa (C) methaphoxide	tris(2-methyl-1-aziridinyl)phosphine oxide	93	156

(continued)

Common name, class,[a] and trade name	Chemical name	Oral LD$_{50}$	Dermal LD$_{50}$
methamidophos (I) Monitor®	O,S-dimethyl phosphoramidothioate	13	110
methidathion (I,M) Supracide®	O,O-dimethyl phosphorodithioate S-ester with 4-(mercapto-methyl)-2-methoxy-Δ²-1,3,4-thiadiazolin-5-one	25	375
methiocarb (I,M) Mesurol®	4-(methylthio)-3,5-xylyl	130	2,000
methiotepa (C) MAPS	tris(2-methyl-1-aziridinyl)phosphine sulfide	—	—
methomyl (I) Lannate®, Nudrin®	S-methyl N-[(methylcarbamoyl)oxy]thioacetimidate	17	1,000
methoprene (GR) Altosid®	isopropyl (E,E)-11-methoxy-3,7,11-trimethyl-2,4-dodecadienoate	34,600	5,000
methotrexate (C) amethopterin	N-[p-[[(2,4-diamino-6-pteridinyl)methyl]methyl-amino]benzoyl]glutamic acid	—	—
methoxychlor (I) Marlate®	1,1,1-trichloro-2,2-bis(p-methoxyphenyl)ethane	5,000	2,820
methyl apholate (C)	2,2,4,4,6,6-hexahydro-2,2,4,4,6,6-hexakis (2-methyl-1-aziridinyl)-1,3,5,2,4,6-triazatriphosphorine	—	—
methyl bromide (F)	bromomethane	200 ppm vapor inhalation	
methyl demeton (I) Metasystox®	mixture of O,O-dimethyl S(and O)-[2-(ethylthio)ethyl]phos-phorothioates	64	—
methyl parathion (I)	O,O-dimethyl O-(p-nitrophenyl) phosphorothioate	9	63
methyl trithion (I)	S-[[(p-chlorophenyl)thio]methyl] O,O-dimethyl phosphorodithioate	98	198
mevinphos (I,M) Phosdrin®	methyl 3-hydroxy-α-crotonate dimethyl phosphate	3	3
mexacarbate (I) Zectran®	4-(dimethylamino)-3,5-xylyl methylcarbamate	15	500
Mirex® (I) Dechlorane®	dodecachlorooctahydro-1,3,4-metheno-1H-cyclobuta [cd]pentalene	235	800
Mobam® (I) Mobil MCA-600	benzo[b] thien-4-yl methylcarbamate	234	6,000
Mocap® (see ethoprop)			
Monitor® (see methamidophos)			
monocrotophos (I,M) Azodrin®	dimethyl phosphate ester with (E)-3-hydroxy-N-methylcrotonamide	21	354
Morestan®	(see oxythioquinox)		
Morocide® (see binapacryl)			
morphothion (I) Ekatin M®	O,O-dimethyl S-(morpholinocarbonylmethyl) phosphorodithioate	100	680
Mylerin® (see busulfan)			
naled (I) Dibrom®	1,2-dibromo-2,2-dichloroethyl dimethyl phosphate	430	1,100
Nemacide (F)	O-(2,4-dichlorophenyl) O,O-diethyl phosphorothioate	270	6,000
Nemacur® (see phenamiphos)			
Nemafume® (see DBCP)			
Nemagon® (see DBCP)			
Niagara NIA 33297 (see FMC-33297			

(continued)

Common name, class,[a] and trade name	Chemical name	Oral LD$_{50}$	Dermal LD$_{50}$
nicotine (I,M)	l-1-methyl-2-(3-pyridyl) pyrrolidine	50	140
omethate (I,M) Folimat®	O,O-dimethyl phosphorothioate S-ester with 2-mercapto-N-methylacetamide	50	1,400
Orthene® (see acephate)			
ovex (I,M) Ovotran®	p-chlorophenyl p-chlorobenzene sulfonate	2,000	200
Ovotran® (see ovex)			
oxamyl (I) Vydate®	methyl N',N'-dimethyl-N-[(methylcarbamoyl)oxy]-1-thiooxamimidate	5	2,960
oxydemetonmethyl (I,M) Meta-Systox-R®	S-[2-(ethylsulfinyl)ethyl]O,O-dimethyl phosphorothioate	65	100
oxythioquinox (I,M) Morestan®	cyclic S,S-(6-methyl-2,3-quinoxalinediyl) dithiocarbonate	2,500	2,000
paradichlorobenzene (see p-dichlorobenzene)			
parathion (I) Niran®	O,O-diethyl O-(p-nitrophenyl) phosphorothioate	3	4
Perthane® (I)	1,1-dichloro-2,2-bis(p-ethylphenyl)ethane	6,600	100
phenamiphos (F) Nemacur®	ethyl-4-(methylthio)-m tolyl isopropylphosphoramidate	8	72
phenthoate (I,M) Cidial®	ethyl mercaptophenylacetate S-ester with O,O-dimethyl phosphorodithioate	200	4,000
phorate (I,M) Thimet®	O,O-diethyl S-[(ethylthio)methyl] phosphorodithioate	1	2
phosalone (I,M) Zolone®	O,O-diethyl S-[(6-chloro-2-oxobenzoxazolin-3-yl)methyl]phosphorodithioate	125	1,500
Phosdrin® (see Mevinphos)			
phosfolan (I) Cyolone®	cyclic ethylene p,p-diethyl phosphonodithio imidocarbonate	9	17
phosmet (I) Imidan®	O,O-dimethyl phosphorodithioate S-ester with N-(mercaptomethyl) phthalimide	147	3,160
phosnichlor (I) Chlorthion®	O-(3-chloro-4-nitrophenyl) O-,O-dimethylphosphorothioate	880	—
phosphamidon (I) Dimecron®	dimethyl phosphate ester with 2-chloro-N,N-diethyl-3-hydroxycrotonamide	15	125
Phostex® (M)	bis(dialkyloxyphosphinothioyl)disulfides (alkyl ratio 75% ethyl, 25% isopropyl)	2,500	2,500
Phosvel® (see leptophos)			
phoxim (I) Baythion®	phenylglyoxylonitrile oxime O,O-diethyl phosphorothioate	1,891	1,126
phthalthrin (see tetramethrin)			
Picfume® (see chloropicrin)			
piperonyl butoxide (S) Butacide®	α-[2-(2-butoxy)ethoxy]-4,5-(methylenedioxy)-2-propyltoluene	7,500	7,500
piprotal (S) Tropital®	piperonal bis[2-(2-butoxyethoxy)ethyl]acetal	4,000	10,000
pirimiphos-ethyl (I) Primicid®	O-[2-(diethylamino)-6-methyl-4-pyrimidinyl]O,O-diethyl phosphorothioate	170	1,000
pirimiphos-methyl (I,M) Actellic®	O-[2-(diethylamino)-6-methyl-4-pyrimidinyl]O,O-dimethyl phosphorothioate	2,050	2,000

(continued)

Common name, class,[a] and trade name	Chemical name	Oral LD$_{50}$	Dermal LD$_{50}$
Plictran® (see cyhexatin)			
Pounce® (see FMC-33297)			
Primicid® (see pirimiphos-ethyl)			
promecarb (I) Minacide®	m-cym-5-yl methylcarbamate	35	—
propargite (M) Omite®	2-(p-tert-butylphenoxy)cyclohexyl 2-propynyl sulfite	2,200	5,000
propoxur (I) Baygon®	o-isopropoxyphenyl methylcarbamate	95	1,000
propylene dichloride (F)	1,2-dichloropropane	>2,000	—
propylene oxide (F)	propylene oxide	3,000 ppm vapor inhalation	
propyl isome (S)	dipropyl 5,6,7,8-tetrahydro-7-methylnaphtho[2,3-d]-1,3-dioxole-5,6-dicarboxylate	15,000	375
Pydrin® (see SD-43775)			
Pyrethrins (I)	mixture of pyrethrins and cinerins	200	1,800
quinalphos (I,M) Bayrusil®	O,O-diethyl O-2-quinoxalinyl phosphorothioate	65	1,200
resmethrin (I) Chryson®	(5-benzyl-3-furyl)methyl cis-trans-(±)-2,2-dimethyl-3-(2-methylpropenyl) cyclopropanecarboxylate	1,500	3,040
d-trans-resmethrin (I) bioresmethrin	(5-benzyl-3-furyl)methyl trans-(±)-2,2-dimethyl-3-(2-methylpropenyl) cyclopropanecarboxylate	—	—
ronnel (I) Korlan®, Trolene®	O,O-dimethyl O-(2,4,5-trichlorophenyl) phosphorothioate	906	1,000
rotenone (I) cube, derris	1,2,12,12a-tetrahydro-2-isopropenyl-8-,9-dimethoxy[1]benzo-pyrano[2,4-b]furo[2,3-b][1]benzopyran-6(6aH)-one	60	>1,000
Ruelene® (see crufomate)			
ryania (I)	Ryania speciosa	750	4,000
sabadilla (I)	Schoenocaulon officinale	4,000	—
schradan (I,M) OMPA	octamethylpyrophosphoramide	5	13
SD-43775 (I) Pydrin®	benzeneacetic acid, 4-chloro-α(1-methylethyl)-cyano(3-phenoxyphenyl) methyl ester	451	1,000
sesamex (S) Sesoxane®	2-(2-ethoxyethoxy)ethyl 3,4-(methylenedioxy)phenyl acetal of acetaldehyde	2,000	11,000
sesamin (S) asarinin	2,6-bis[3,4-(methylenedioxy)phenyl]-3,7-dioxabicyclo[3.3.0]octane	2,000	11,000
sesamolin (S)	6-[3,4-(methylenedioxy)phenoxy]-2-[3,4-methylenedioxy)-phenyl]-3,7-dioxabicyclo[3.3.0]octane	2,000	10,000
Sevin® (see carbaryl)			
Stauffer N-2596 (I) BAY 36743	S-(p-chlorophenyl) O-ethyl ethylphosphonodithioate	4	261
stirofos (I) Gardona®, Rabon®	2-chloro-1-(2,4,5-trichlorophenyl)vinyl dimethylphosphate	4,000	5,000
Strobane T® (I) toxaphene	terpene polychlorinates (65% chlorine)	40	600
sulfotepp (I,M) thiotepp	O,O,O',O'-tetraethyl dithiopyrophosphate	5	—
sulfoxide (S) Sulfox-Cide®	1,2-(methylenedioxy)-4-[2-(octylsulfinyl)propyl]benzene	2,000	9,000

(continued)

Common name, class,[a] and trade name	Chemical name	Oral LD_{50}	Dermal LD_{50}
sulfuryl fluoride (F,I) Vikane®	sulfuryl fluoride		
Sulphenone® (M)	p-chlorophenyl phenyl sulfone	1,400	—
Sumithion® (see fenitrothion)			
Supona® (see chlorfenvinphos)			
Supracide® (see methidathion)			
Systox® (see demeton)			
Tactic® (see amitraz)			
Tartan® (see cyanthoate)			
TDE (I) DDD	1,1-dichloro-2,2-bis(p-chlorophenyl) ethane	3,400	4,000
Tedion® (see tetradifon)			
temephos (I) Abate®	O,O'-(thiodi-p-phenylene) O,O,O',O',tetramethyl phosphorothioate	1,000	4,000
tepa (C) aphoxide	tris(1-aziridinyl)phosphine oxide	37	—
tepp (I)	tetraethyl pyrophosphate	0.2	2
terbufos (I) Counter®	S-[(tert-butylthio)methyl] O,O-diethyl phosphorodithioate	4	1
tetradifon (M) Tedion®	p-chlorophenyl 2,4,5-trichlorophenyl sulfone	5,000	1,000
tetramethrin (I) Neo-Pynamin®	1-cyclohexane-1,2-dicarboximidomethyl 2,2-dimethyl-3-(2-methylpropenyl) cyclopropanecarboxylate	20,000	15,000
tetrasul (M) Animert®	p-chlorophenyl 2,4,5-trichlorophenyl sulfide	3,960	2,000
Thanite® (I)	isobornyl thiocyanoacetate	1,000	6,000
Thimet® (see phorate)			
thiofanox (I,M) Diamond-Shamrock DS-15647	3,3-dimethyl-1-(methylthio)-2-butanone O-[(methyl-amino)carbonyl]oxime	9	39
thioquinox (M) Eradex®	cyclic 2,3-quinoxalinediyl trithiocarbonate	3,400	—
Thuracide® (see Bacillus thuringiensis)			
Torak® (see dialifor)			
toxaphene (I)	chlorinated camphene containing 67–69% chlorine	40	600
tretamine (C) TEM	2,4,6-tris(1-aziridinyl)s-triazine	1	—
triazophos (I,M) Hostathion®	O,O-diethyl O-(1-phenyl-1H-1,2,4-triazol-3-yl) phosphorothioate	82	1,100
trichlorfon (I) Dipterex®, Dylox®, Neguvon®	dimethyl (2,2,2-trichloro-1-hydroxyethyl)phosphonate	450	2,000
trichloronat (I) Agritox®	O-ethyl O-[2-4-5-trichlorophenyl) ethylphosphonothioate	16	135
Trithion® (see carbophenothion)			
U-36059 (see amitraz)			
Vendex® (M) Shell SD-14114	hexakis (2-methyl-2-phenylpropyl)distannoxane	2,000	2,000
Viron/H® (see Heliothis virus)			
Zectran® (see mexacarbate)			
Zinophos® (I) Nemaphos®	O,O-diethyl O-pyrazinyl phosphorothioate	9	8
Zolone® (see phosalone)			

[a] Classes are (C) Insect chemosterilant, (I) Insecticide, (M) Acaricide, (S) Insecticide synergist, (F) Fumigant, and (GR) Growth regulator.
[b] (—)-Toxicity data unavailable from standard sources.

Common and chemical names of herbicides and their oral LD$_{50}$s to rats.[a]

Common name or designation	Chemical name	LD$_{50}$ (mg/kg)
acrolein	2-propenal	46
alachlor	2-chloro-2',6'-diethyl-N-(methoxymethyl)acetanilide	1,200
ametryn	2-(ethylamino)-4-(isopropylamino)-6-(methylthio)-s-triazine	1,110
amitrole	3-amino-s-triazole	1,100
AMS	ammonium sulfamate	3,900
asulam	methyl sulfanilylcarbamate	2,000
atraton	2-(ethylamino)-4-(isopropylamino)-6-methoxy-s-triazine	1,465
atrazine	2-chloro-4-(ethylamino)-6-(isopropylamino)-s-triazine	3,080
barban	4-chloro-2-butynyl m-chlorocarbanilate	600
benazolin	4-chloro-2-oxobenzothiazolin-3-ylacetic acid	3,000
benefin	N-butyl-N-ethyl-α,α,α-trifluoro-2,6-dinitro-p-toluidine	10,000
bensulide	O,O-diisopropyl phosphorodithioate S-ester with N-(2-mercaptoethyl)benzenesulfonamide	770
bentazon	3-isoropyl-1H-2,1,3-benzothiadiazin-(4) 3H-one 2,2-dioxide	1,100
benzadox	(benzamidooxy)acetic acid	5,600
benzipram	3,5-dimethyl-N-(1-methylethyl-N-(phenylmethyl) benzamide	—
bifenox	methyl 5-(2,4-dichlorophenoxy)-2-nitrobenzoate	6,400
bromacil	5-bromo-3-sec-butyl-6-methyluracil	5,200
bromoxynil	3,5-dibromo-4-hydroxybenzonitrile	190
butachlor	N-(butoxymethyl)-2-chloro-2',6'-diethylacetanilide	1,200
butam	2,2-dimethyl-N-(1-methylethyl)-N-(phenylmethyl)propanamide	—
buthidazole	3-[5-(1,1-dimethylethyl)-1,3,4-thiadiazol-2-yl]-4-hydroxy-1-methyl-2-imidazolidinone	—
butralin	4-(1,1-dimethylethyl)-N-(1-methylpropyl)-2,6-dinitrobenzenamine	1,000
butylate	S-ethyl diisobutylthiocarbamate	4,000
cacodylic acid	hydroxydimethylarsine oxide	830
carbetamide	D-N-ethylacetamide carbanilate (ester)	11,000
CDAA	N-N-diallyl-2-chloroacetamide	700
CDEC	2-chloroallyl diethyldithiocarbamate	850
chloramben	3-amino-2,5-dichlorobenzoic acid	3,500
chlorazine	2-chloro-4,6-bis (diethylamino)-s-triazine	850

(continued)

Common name or designation	Chemical name	LD_{50} (mg/kg)
chlorbromuron	3-(4-bromo-3-chlorophenyl)-1-methoxy-1-methylurea	2,150
chloroxuron	3-[p-(p-chlorophenoxy)phenyl]-1,1-dimethylurea	3,000
chlorpropham	isopropyl m-chlorocarbanilate	3,800
cisanilide	cis-2,5-dimethyl-N-phenyl-1-pyrrolidinecarboxamide	4,100
CMA	calcium methanearsonate	4,000
cyanazine	2-[[4-chloro-6-(ethylamino)-s-triazin-2-yl]amino]-2-methylpropionitrile	149
cycloate	S-ethyl N-ethylthiocyclohexanecarbamate	3,160
cycluron	3-cyclooctyl-1,1-dimethylurea	4,600
cyperquat	1-methyl-4-phenylpyridinium	35
cyprazine	2-chloro-4-(cyclopropylamino)-6-(isopropylamino)-s-triazine	1,200
cyprazole	N-[5-(2-chloro-1,1-dimethylethyl)-1,3,4-thiadiazol-2-yl]cyclopropanecarboxamide	—
cypromid	3',4'-dichlorocyclopropanecarboxanilide	—
dalapon	2,2-dichloropropionic acid	6,500
dazomet	tetrahydro-3,5-dimethyl-2H-1,3,5-thiadiazine-2-thione	650
DCPA	dimethyl tetrachloroterephthalate	3,000
desmedipham	ethyl M-hydroxycarbanilate carbanilate (ester)	8,000
desmetryn	2-(isopropylamino)-4-(methylamino)-6-(methylthio)-s-triazine	1,390
diallate	S-(2,3-dichloroallyl)diisopropylthiocarbamate	395
dicamba	3,6-dichloro-o-anisic acid	1,040
dichlobenil	2,6-dichlorobenzonitrile	3,160
dichlone	2,3-dichloro-1,4-naphthoquinone	1,300
dichlorprop	2-(2,4-dichlorophenoxy)propionic acid	400
difenzoquat	1,2-dimethyl-3,5-diphenyl-1H-pyrazolium	470
dinitramine	N^4,N^4-diethyl-α,α,α-trifluoro-3,5-dinitrotoluene-2,4-diamine	3,000
dinoseb	2-sec-butyl-4,6-dinitrophenol	58
diphenamid	N,N-dimethyl-2,2-diphenylacetamide	1,000
dipropetryn	2-ethylthio-4-6-bis-isopropylamino-s-triazine	4,050
diquat	6,7-dihydrodipyrido[1,2-α:2',1'-c]pyrazinediium ion	231
diuron	3-(3,4-dichlorophenyl)-1,1-dimethylurea	3,400
DNOC	4,6-dinitro-o-cresol	30
DSMA	disodium methanearsonate	600
endothall	7-oxabicyclo[2.2.1]heptane-2,3-dicarboxylic acid	38
EPTC	S-ethyldipropylthiocarbamate	1,367
erbon	2-(2,4,5-trichlorophenoxy)ethyl 2,2-dichloropropionate	1,000
ethalfluralin	N-ethyl-N-(2-methyl-2-propenyl)-2,6-dinitro-4-(trifluoromethyl)benzenamine	10,000
ethiolate	S-ethyl diethylthiocarbamate	710
fenac	(2,3,6-trichlorophenyl)acetic acid	1,780
fenuron	1,1-dimethyl-3-phenylurea	6,400
fenuron TCA	1,1-dimethyl-3-phenylurea mono(trichloroacetate)	5,700

(continued)

Common name or designation	Chemical name	LD_{50} (mg/kg)
fluchloralin	N-(2-chloroethyl)-2,6-dinitro-N-propyl-4-(trifluoromethyl)aniline	1,550
fluometuron	1,1-dimethyl-3-(α,α,α-trifluoro-m-tolyl)urea	7,880
fluorodifen	p-nitrophenyl α,α,α-trifluoro-2-nitro-p-tolyl ether	15,000
glyphosate	N-(phosphonomethyl)glycine	4,320
hexaflurate	potassium hexafluoroarsenate	1,200
ioxynil	4-hydroxy-3,5-diiodobenzonitrile	110
ipazine	2-chloro-4-(diethylamino)-6-(isopropylamino)-S-triazine	—
isocil	5-bromo-3-isopropyl-6-methyluracil	5,200
isopropalin	2,6-dinitro-N,N-dipropylcumidine	5,000
karbutilate	tert-butylcarbamic acid ester with 3(m-hydroxyphenyl)-1,1-dimethylurea	3,000
lenacil	3-cyclohexyl-6,7-dihydro-1H-cyclopentapyrimidine-2,4(3H,5H)-dione	11,000
linuron	3-(3,4-dichlorophenyl)-1-methoxy-1-methylurea	1,500
MAA	methanearsonic acid	1,300
MAMA	monoammonium methanearsonate	750
MCPA	[(4-chloro-o-tolyl)oxy]acetic acid	700
MCPB	4-[(4-chloro-o-tolyl)oxy]butyric acid	680
mecoprop	2-[(4-chloro-o-tolyl)oxy]propionic acid	930
mefluidide	N-[2,4-dimethyl-5][(trifluoromethyl)sulfonyl]-amino[phenyl]acetamide	—
metham	sodium methyldithiocarbamate	820
methazole	2-(3,4-dichlorophenyl)-4-methyl-1,2,4-oxadiazolidine-3,5-dione	1,350
metho-bromuron	3-(p-bromophenyl)-1-methoxy-1-methylurea	2,700
metribuzin	4-amino-6-tert-butyl-3-(methylthio)-as-triazine-5(4H)one	1,937
MH	1,2-dihydro-3,6-pyridazinedione	2,200
molinate	S-ethyl hexahydro-1H-azepine-1-carbothioate	584
monolinuron	3-(p-chlorophenyl)-1-methoxy-1-methylurea	2,250
monuron	3-(p-chlorophenyl)-1,1-dimethylurea	3,600
monuron TCA	3-(p-chlorophenyl)-1,1-dimethylurea mono(trichloroacetate)	2,300
MSMA	monosodium methanearsonate	700
napropamide	2-(α-naphthoxy)-N,N-diethylpropionamide	5,000
naptalam	N-1-naphthylphthalamic acid	1,770
neburon	1-butyl-3-(3,4-dichlorophenyl)-1-methylurea	11,000
nitralin	4-(methylsulfonyl)-2,6-dinitro-N,N-dipropylaniline	2,000
nitrofen	2,4-dichlorophenyl-p-nitrophenyl ether	2,630
nitrofluorfen	2-chloro-1-(4-nitrophenoxy)-4-(trifluoromethyl)benzene	—
norea	3-(hexahydro-4,7-methanoindan-5-yl)-1,1-dimethylurea	2,000
norflurazon	4-chloro-5-(methylamino)-2-(α,α,α-trifluro-m-tolyl)-3(2H)-pyridazinone	8,000
oryzalin	3,5-dinitro-N^4,N^4-dipropylsulfanilamide	10,000
oxadiazon	2-tert-butyl-4-(2,4-dichloro-5-isopropoxyphenyl)-Δ2-1,3,4-oxadiazolin-5-one	3,500

(continued)

Common name or designation	Chemical name	LD_{50} (mg/kg)
oxyfluorfen	2-chloro-1-(3-ethoxy-4-nitrophenoxy)-4-(trifluoromethyl)benzene	—
paraquat	1,1'-dimethyl-4,4'-bipyridinium ion	150
PBA	chlorinated benzoic acid	1,140
PCP	pentachlorophenol	210
pebulate	S-propyl butylethylthiocarbamate	1,120
penoxalin	N-(1-ethylpropyl)-3,4-dimethyl-2,6-dinitrobenzenamine	1,250
perfluidone	1,1,1-trifluoro-N-[2-methyl-4-(phenylsulfonyl)phenyl] methanesulfonamide	633
phenmedipham	methyl m-hydroxycarbanilate m-methylcarbanilate	8,000
picloram	4-amino-3,5,6-trichloropicolinic acid	8,200
procyazine	2-[[4-chloro-6-(cyclopropylamino)-1,3,5-triazine-2-yl]amino]-2-methylpropanenitrile	290
profluralin	N-(cyclopropylmethyl)-α,α,α-trifluoro-2,6-dinitro-N-propyl-p-toluidine	2,200
prometone	2,4-bis(isopropylamino)-6-methoxy-s-triazine	2,980
prometryn	2,4-bis(isopropylamino)-6-(methylthio)-s-triazine	3,150
pronamide	3,5-dichloro(N-1,1-dimethyl-2-propynyl) benzamide	5,620
propachlor	2-chloro-N-isopropylacetanilide	710
propanil	3'-4'-dichloropropionalide	1,384
propazine	2-chloro-4,6-bis(isopropylamino)-s-triazine	5,000
propham	isopropyl carbanilate	9,000
prosulfalin	N-[[4-(dipropylamino)-3,5-dinitrophenyl]sulfonyl]-S,S-dimethylsulfilimine	2,000
prynachlor	2-chloro-N-(1-methyl-2-propynyl)acetanilide	1,177
pyrazon	5-amino-4-chloro-2-phenyl-3(2H)-pyridazinone	3,300
secbumeton	N-ethyl-6-methoxy-N'(1-methylpropyl)-1,3,5-triazine-2,4-diamine	2,680
sesone	2-(2,4-dichlorophenoxy)ethyl sodium sulfate	1,400
siduron	1-(2-methylcyclohexyl)-3-phenylurea	7,500
silvex	2-(2,4,5-trichlorophenoxy)propionic acid	375
simazine	2-chloro-4,6-bis(ethylamino)-s-triazine	5,000
simeton	2,4-bis(ethylamino)-6-methoxy-s-triazine	535
simetryn	2,4-bis(ethylamino)-6-(methylthio)-s-triazine	1,830
sodium arsenite	sodium arsenite	10–50
sodium chlorate	sodium chlorate	1,200
solan	3'-chloro-2-methyl-p-valerotoluidide	10,000
swep	methyl 3,4-dichlorocarbanilate	522
TCA	trichloroacetic acid	3,200
tebuthiuron	N-[5-(1,1-dimethylethyl)-1,3,4-thiadiazol-2-yl]-N,N'dimethylurea	600
terbacil	3-tert-butyl-5-chloro-6-methyluracil	5,000
terbuthylazine	2-(tert-butylamino)-4-chloro-6-(ethylamino-s-triazine	2,160
terbutol	2,6-di-tert-butyl-p-tolyl methylcarbamate	34,000
terbutryn	2-(tert-butylamino)-4-(ethylamino)-6-(methylthio)-s-triazine	2,100

(continued)

Common name or designation	Chemical name	LD_{50} (mg/kg)
triallate	S-(2,3,3-trichloroallyl)diisopropylthiocarbamate	1,675
triclopyr	[(3,5,6-trichloro-2-pyridinyl)oxy]acetic acid	713
tricamba	3,5,6-trichloro-o-anisic acid	970
trietazine	2-chloro-4-(diethylamino)-6-(ethylamino)-s-triazine	2,830
trifluralin	α,α,α-trifluoro-2,6-dinitro-N,N-dipropyl-p-toluidine	3,700
trimeturon	1-(p-chlorophenyl)-2,3,3-trimethylpseudourea	—
2,3,6-TBA	2,3,6-trichlorobenzoic acid	750
2,4-D	(2,4-dichlorophenoxy)acetic acid	375
2,4,DB	4-(2,4-dichlorophenoxy)butyric acid	500
2,4-DEP	tris[2-(2,4-dichlorophenoxy)ethyl]phosphite	850
2,4,5-T	(2,4,5-trichlorophenoxy)acetic acid	300
vernolate	S-propyl dipropylthiocarbamate	1,710

[a] Proprietary or trade names are not shown due to their abundance and transient nature in the pesticide trade. Dashes indicate that LD_{50} is not available.

APPENDIX C

Common, trade, and chemical names of fungicides and bactericides and their oral LD$_{50}$s to rats.[a]

Common name and trade name	Chemical name	LD$_{50}$
anilazine, Dyrene®	2,4-dichloro-6-(o-chloroanilino)-s-triazine	2,710
basic copper sulfate	basic copper sulfate	nontoxic
benomyl, Benlate®	methyl-1-(butylcarbamoyl)-2-benzamidazole carbamate	9,590
binapacryl, Morocide®	2-sec-butyl-4,6-dinitrophenyl 3-methyl-2-butenoate	136
biphenyl, Diphenyl®	diphenyl	3,280
Bordeaux	mixture of copper sulfate and calcium hydroxide forming basic copper sulfates	nontoxic
Busan-72®	2-(thiocyanomethylthio)benzothiazole	1,590
butazon	3-isopropyl-1H-2,1,3-benzothiadiazin 4(3H)-one 2,2-dioxide	—
Cadminate	cadmium succinate	660
cadmium chloride	cadmium chloride	medium toxicity
captafol, Difolatan®	cis-N(1,1,3,2-tetrachloroethyl)thio)-4-cyclohexene-1,2-dicarboximide	6,200
captan	N-(trichloromethylthio)-4-cyclohexene-1,2-dicarboximide	9,000
carboxin, DCMO, Vitavax®	5,6-dihydro-2-methyl-1-4-oxathiin-3-carboxanilide	3,820
ceresan, MEMC	2-methoxyethylmercuric chloride	22
chloroneb, Demosan®	1,4-dichloro-2,5-dimethoxybenzene	11,000
chlorothalonil, Daconil® 2787, Bravo®	tetrachloro isophthalonitrile	10,000
copper ammonium carbonate	copper ammonium complex	low
copper hydroxide	copper hydroxide	low
copper naphthenates	cupric cyclopentanecarboxylate	low
copper oxychloride	basic copper chloride	low
copper oxychloride sulfate	mixture of basic copper chloride and basic copper sulfate	nontoxic
copper quinolinolate	copper-8-quinolinolate	10,000
copper salts of organic acids		low
copper sulfate bluestone	copper sulfate pentahydrate	300
copper sulfate monohydrate	cupric sulfate monohydrate	nontoxic
cyclohexamide, Acti-Dione®	β(2-(3,5-dimethyl-2-oxocyclohexyl)-2-hydroxyethyl)-glutarimide	2
dichlone, Phygon®	2,3-dichloro-1,4-naphthoquinone	1,300
dichlozoline, Sclex®	3-(3,5-dichlorophenyl)-5,5-dimethyl-2,4-oxazolidinedione	3,000
dicloran, Botran®	2,6-dichloro-4-nitroaniline	1,500

(continued)

Common name and trade name	Chemical name	LD_{50}
dinocap, Karathane®, Mildex®	mixture of 2,4-dinitro-6-octylphenyl crotonate and 2,6-dinitro-4-octylphenyl crotonate	980
ditalimfos, Dowco 199®	O,O-diethyl phthalimidophosphonothioate	5,660
dodine, Cyprex®, Melprex®	n-dodecylguanidine acetate	660
Dowcide A®	sodium-o-phenyl phenate	1,160
ethazol, Terrazole®	5-ethoxy-3-trichloromethyl-1,2,4-thiadiazole	2,000
fentin acetate, Brestan®, Fentin®	triphenyltin acetate	125
fentin hydroxide	triphenyltin hydroxide	108
ferbam	ferric dimethyldithiocarbamate	1,000
folpet, Phaltan®	N-(trichloromethylthio)-phthalimide	10,000
Frucote®	2-aminobutane	380
hexachlorobenzene	1,2,3,4,5,6-hexachlorobenzene	10,000
Indar®	4-N-butyl-1,2,4-triazole	50
lime sulfur	calcium polysulfide	skin irritation
maneb	manganese ethylenebisdithiocarbamate	6,750
metiram, Polyram®	mixture of ammoniates of ethylenebis (dithiocarbamate)- zinc and ethylenebisdithiocarbamic acid cyclic anhydrosulfides and disulfides	6,200
milneb, thiadiazin	3,3'-ethylenebis (tetrahydro-4,6-dimethyl-2H-1,3,5-thiadiazine-2-thione	5,000
nabam	disodium ethylenebisdithiocarbamate	395
oxycarboxin, DCMOD, Plantvax®	5,6-dihydro-2-methyl-1,4-oxathiin-3-carboxanilide-4,4-dioxide	2,000
oxythioquinox, Morestan®	6-methyl-1-3-dithiolo (4,5-b)quinoxalin-2-one	2,500
parinol, Parnon®	α,α-bis(p-chlorophenyl)-3-pyridine-methanol	5,000
PCNB	pentachloronitrobenzene	12,000
PCP	pentachlorophenol	210
phenyl phenol, Dowcide 1®	o-phenyl phenol	2,700
piperalin, Pipron®	3-(2-methylpiperidino)propyl 3,4-dichlorobenzoate	2,500
PMA (many trade names)	phenylmercury acetate	25
propineb, Antracol®	zinc-propylenebisdithiocarbamate	8,500
prothiocarb, Previcure®	S-ethyl-N-(3-dimethylaminopropyl)thiol carbamate hydrochloride	1,300
pyroxychlor, Dowco 269®	2-chloro-6-methoxy-4-(trichloromethyl) pyridine	1,500
streptomycin	2,4-diguanidino-3,5,6-trihydroxycyclohexyl-5-deoxy-2-O-2-deoxy-2-methylamino-α-glucopyranosyl)-3-formyl pentofuranoside	9,000
sulfur	elemental sulfur in many formulations	nontoxic
TEC, Tecoram®	bis-(dimethyldithiocarbamoyl)-ethylenebisdithiocarbamate	3,000
thiabendazole, TBZ	2-(4'-thiazoyl)benzimidazole	3,100
thiophanate, Topsin®	1,2-bis(3-ethoxycarbonyl-2-thioureido) benzene	15,000

(continued)

Common name and trade name	Chemical name	LD_{50}
thiram, Arasan®, TMTD	tetramethylthiuramidisulfide	780
tricyclazole, EL-291	5-methyl-1,2,4-triazole(3,4-b) benzothiazole	250
triflorine	N,N'-(piperazinediylbis(2,2,2-trichloro-ethylidene))bis(formamide)	6,000
urbacide	methylarsenic-dimethyl-dithiocarbamate	175
zinc coposil	basic cupric zinc sulfate complex	nontoxic
zineb, Dithane® Z-78	zinc ethylenebis(dithiocarbamate)	5,200
ziram	zinc dimethyldithiocarbamate	1,400

[a] Dashes indicate LD_{50} not available.

Common, trade, and chemical names of rodenticides and their oral LD_{50}s to rats.

Common name and trade name	Chemical name	LD_{50}
alphachloralose Glucochloralose®	alpha-chloralose	400
antu, Antu®	α-naphthylthiourea	6
barium carbonate	barium carbonate	630
chlorophacinone	2-((p-chlorophenyl)phenylacetyl)-1,3-indandione	20.5
Compound 1080, Ten-Eighty	sodium fluoroacetate, or sodium monofluoroacetate	0.22
Compound 1081 Ten-Eighty-One	fluoracetamide	5.75
coumachlor, Tomorin®, Ratilan®	3-(α-acetonyl-4-chlorobenzyl)-4-hydroxycoumarin	900
coumafuryl, Fumarin®	3-(α-acetonylfurfuryl)-4-hydroxycoumarin	25
coumatetralyl, Racumin®	3-(α-tetralyl)-4-hydroxycoumarin	16.5
crimidine, Castrix®	2-chloro-4-dimethylamino-6 methylpyrimidine	1.25
dicumarol, Dicoumarin®	3,3'-methylene-bis-(4-hydroxycoumarin)	541
diphacinone, Diphacin®	2-diphenylacetyl-1,3-indandione	3
endrin	1,2,3,4,10,10-hexachloro-7-epoxy-1,4,4a,5,6,7,8,8a-octahydro exo-1,4-exo-5,8-dimethanonaphthalene	10
Gophacide®	O,O-bis(4-chlorophenyl)acetimidoylphosphoramidothioate	3.7
norbormide, Radicate®	5-(α-hydroxy-α-2-pyridyl-benzyl)-7-(α-2-pyridyl-benzylidene)-5-norbornene-2,3-dicarboximide	5.3
pindone, Pival®	2-pivaloylindane-1,3-dione	280
Prolin®	warfarin plus sulfaquinoxaline	1000
red squill	chemical structure undetermined	0.7
sodium arsenite	sodium arsenite	10
sodium cyanide, Cymag®	sodium cyanide	0.5–2.0
strychnine	alkaloid from tree, Strychnos nux-vomica	1 to 30
thallium sulfate, Ratox®	thallium sulfate	16
Vacor®, RH-787	N-3-pyridylmethyl N'-nitrophenyl urea	12.3
Valone®	2-isovaleryl-1,3-indandione	50
warfarin, d-Con®, Warf®	3-(α-acetonylbenzyl)-4-hydroxycoumarin	186
yellow phosphorus	elemental phosphorus	<6
zinc phosphide, Phosvin®	zinc phosphide	45

Glossary

AAPCO. Association of American Pesticide Control Officials, Inc.

Abscission. Process by which a leaf or other part is separated from the plant.

Absorption. Process by which pesticides are taken into tissues, namely plants, by roots or foliage (stomata, cuticle, etc.).

Acaricide (miticide). An agent that destroys mites and ticks.

Acetylcholine (ACh). Chemical transmitter of nerve and nerve-muscle impulses in animals.

Activator. Material added to a fungicide to increase toxicity.

Active ingredient (a.i. or AI). Chemicals in a product that are responsible for the pesticidal effect.

Acute toxicity. The toxicity of a material determined at the end of 24 hours; to cause injury or death from a single dose or exposure.

Adjuvant. An ingredient that improves the properties of a pesticide formulation. Includes wetting agents, spreaders, emulsifiers, dispersing agents, foam suppressants, penetrants, and correctives.

Adsorption. Chemical and/or physical attraction of a substance to a surface. Refers to gases, dissolved substances, or liquids on the surface of solids or liquids.

Aerosol. Colloidal suspension of solids or liquids in air.

Adulterated pesticide. A pesticide that does not conform to the professed standard or quality as documented on its label or labeling.

Agroecosystem. An agricultural area sufficiently large to permit long-term interactions of all the living organisms and their nonliving environment.

Algicide. Chemical used to control algae and aquatic weeds.

Alkylating agent. Highly active compounds (chemosterilants) that replace hydrogen atoms with alkyl groups, usually in cells undergoing division.

Annual. Plant that completes its life cycle in one year, i.e., germinates from seed, produces seed, and dies in the same season.

Antagonism. Decreased activity arising from the effect of one chemical or another (opposite of synergism).

Antibiotic. Chemical substance produced by a microorganism and that is toxic to other microorganisms.

Anticoagulant. A chemical that prevents normal bloodclotting. The active ingredient in some rodenticides.

Antidote. A practical treatment, including first aid, used in the treatment of pesticide poisoning or some other poison in the body.

Antimetabolite. Chemicals that are structurally similar to biologically active metabolites, and that may take their place detrimentally in a biological reaction.

Antitranspirant. A chemical applied directly to a plant that reduces the rate of transpiration or water loss by the plant.

Apiculture. Pertaining to the care and culture of bees.

Aromatics. Solvents containing benzene or compounds derived from benzene.

Atropine (atropine sulfate). An antidote used to treat organophosphate and carbamate poisoning.

Attractant, insect. A substance that lures insects to trap or poison-bait stations. Usually classed as food, oviposition, and sex attractants.

Auxin. Substance found in plants that stimulates cell growth in plant tissues.

Avicide. Lethal agent used to destroy birds but also refers to materials used for repelling birds.

Aziridine. Chemical classification of chemosterilants containing three-membered rings composed of one nitrogen and two carbon atoms.

Bactericide. Any bacteria-killing chemical.

Bacteriostat. Material used to prevent growth or multiplication of bacteria.

Band application. Application to a continuous restricted band such as in or along a crop row, rather than over the entire field area.

Biennial. Plant that completes its growth in 2 years. The first year it produces leaves and stores food; the second year it produces fruit and seeds.

Biological control agent. Any biological agent that adversely affects pest species.

Biomagnification. The increase in concentration of a pollutant in animals as related to their position in a food chain, usually referring to the persistent, organochlorine insecticides and their metabolites.

Biota. Animals and plants of a given habitat.

Biotic insecticide. Usually microorganisms known as *insect pathogens* that are applied in the same manner as conventional insecticides to control pest species.

Biotype. Subgroup within a species differing in some respect from the species such as a subgroup that is capable of reproducing on a resistant variety.

Botanical pesticide. A pesticide produced from naturally occurring chemicals found in some plants. Examples are nicotine, pyrethrum, strychnine, and rotenone.

Brand. The name, number, or designation of a pesticide.

Broadcast application. Application over an entire area rather than only on rows, beds, or middles.

Broad-spectrum insecticide. Nonselective, having about the same toxicity to most insects.

Calibrate. To determine the amount of pesticide that will be applied to the target area.

Carbamate insecticide. One of a class of insecticides derived from carbamic acid.

Carcinogen. A substance that causes cancer in animal tissue.

Carrier. An inert material that serves as a diluent or vehicle for the active ingredient or toxicant.

Causal organism. The organism (pathogen) that produces a given disease.

Certified applicator. Commercial or private person qualified to apply restricted-use pesticides as defined by the EPA.

Chelating agent. Certain organic chemicals (i.e., ethylenediaminetetraacetic acid) that combine with metal to form soluble chelates and prevent conversion to insoluble compounds.

Chemical name. Scientific name of the active ingredient(s) found in the formulated product. The name is derived from the chemical structure of the active ingredient.

Chemosterilant. Chemical compounds that cause sterilization or prevent effective reproduction.

Chemotherapy. Treatment of a diseased organism, usually plants, with chemicals to destroy or inactivate a pathogen without seriously affecting the host.

Chemtrec. A toll-free, long-distance, telephone service that provides 24-hour emergency pesticide information (800-424-9300).

Chlorosis. Loss of green color in foliage.

Cholinesterase (ChE). An enzyme of the body necessary for proper nerve function that is inhibited or damaged by organophosphate or carbamate insecticides taken into the body by any route.

Chronic toxicity. The toxicity of a material determined beyond 24 hours and usually after several weeks of exposure.

Common pesticide name. A common chemical name given to a pesticide by a recognized committee on pesticide nomenclature. Many pesticides are known by a number of trade or brand names but have only one recognized common name. For example, the common name for Sevin insecticide is *carbaryl*.

Compatible. *Compatibility.* When two materials can be mixed together with neither affecting the action of the other.

Concentration. Content of a pesticide in a liquid or dust; for example, pounds/gallon or percent by weight.

Contact herbicide. Phytotoxin by contact with plant tissue rather than as a result of translocation.

Contamination. The presence of an unwanted pesticide or other material in or on a plant, animal, or their by-products; soil; water; air; structure; etc. (See *Residue*).

Cumulative pesticides. Those chemicals that tend to accumulate or build up in the tissues of animals or in the environment (soil, water).

Curative pesticide. A pesticide that can inhibit or eradicate a disease-causing organism after it has become established in the plant or animal.

Cutaneous toxicity. Same as dermal toxicity.

Cuticle. Outer covering of insects.

Deciduous. Plants that lose their leaves during the winter.

Decontaminate. The removal or breakdown of any pesticide chemical from any surface or piece of equipment.

Deflocculating agent. Material added to a spray preparation to prevent aggregation or sedimentation of the solid particles.

Defoliant. A chemical that initiates abscission.

Deposit. Quantity of a pesticide deposited on a unit area.

Dermal toxicity. Toxicity of a material as tested on the skin, usually on the shaved belly of a rabbit; the property of a pesticide to poison an animal or human when absorbed through the skin.

Desiccant. A chemical that induces rapid desiccation of a leaf or plant part.

Desiccation. Accelerated drying of plant or plant parts.

Detoxify. To make an active ingredient in a pesticide or other poisonous chemical harmless and incapable of being toxic to plants and animals.

Diluent. Component of a dust or spray that dilutes the active ingredient.

Disinfectant. A chemical or other agent that kills or inactivates disease-producing microorganisms in animals, seeds, or other plant parts. Also commonly referrs to chemicals used to clean or surface sterilize inanimate objects.

DNA. Deoxyribonucleic acid.

Dormant spray. Chemical applied in winter or very early spring before treated plants have started active growth.

Dose, dosage. Same as rate. The amount of toxicant given or applied per unit of plant, animal, or surface.

Drift, spray. Movement of airborne spray droplets from the spray nozzle beyond the intended contact area.

EC_{50}. The median effective concentration (ppm or ppb) of the toxicant in the environment (usually water) that produces a designated effect in 50 percent of the test organisms exposed.

Ecdysone. Hormone secreted by insects essential to the process of molting from one stage to the next.

Ecology. Derived from the Greek *oikos*, "house or place to live." A branch of biology concerned with organisms and their relation to the environment.

Economic level. The insect pest level at which additional management practices must be employed to prevent economic losses.

Ecosystem. The interacting system of all the living organisms of an area and their nonliving environment.

ED_{50}. The median effective dose, expressed as mg/kg of body weight, which produces a designated effect in 50 percent of the test organisms exposed.

Emulsifiable concentrate. Concentrated pesticide formulation containing organic solvent and emulsifier to facilitate emulsification with water.

Emulsifier. Surface active substances used to stabilize suspensions of one liquid in another; for example, oil in water.

Emulsion. Suspension of miniscule droplets of one liquid in another.

Environment. All the organic and inorganic features that surround and affect a particular organism or group of organisms.

Environmental Protection Agency (EPA). The federal agency responsible for pesticide rules and regulations, and all pesticide registrations.

EPA. The Environmental Protection Agency.

EPA Establishment Number. A number assigned to each pesticide production plant by EPA. The number indicates the plant at which the pesticide product was produced and must appear on all labels of that product.

EPA Registration Number. A number assigned to a pesticide product by EPA when the product is registered by the manufacturer or his designated agent. The number must appear on all labels for a particular product.

Eradicant. Applies to fungicides in which a chemical is used to eliminate a pathogen from its host or environment.

Exterminate. Often used to imply the complete extinction of a species over a large continuous area such as an island or a continent.

FEPCA. The Federal Environmental Pesticide Control Act of 1972.

Field scout. A person who samples fields for insect infestations.

FIFRA. The Federal Insecticide, Fungicide and Rodenticide Act of 1947.

Filler. Diluent in powder form.

Fixed coppers. Insoluble copper fungicides where the copper is in a combined form. Usually finely divided, relatively insoluble powders.

Flowable. A type of pesticide formulation in which a very finely ground solid particle is mixed in a liquid carrier.

Foaming agent. A chemical that causes a pesticide preparation to produce a thick foam. This aids in reducing drift.

Fog treatment. The application of a pesticide as a fine mist for the control of pests.

Food chain. Sequence of species within a community, each member of which serves as food for the species next higher in the chain.

Formamidine insecticide. A new group with a new mode of action highly effective against insect eggs and mites.

Formulation. Way in which basic pesticide is prepared for practical use. Includes preparation as wettable powder, granular, emulsifiable concentrate, etc.

Full-coverage spray. Applied thoroughly over the crop to a point of runoff or drip.

Fumigant. A volatile material that forms vapors that destroy insects, pathogens, and other pests.

Fungicide. A chemical that kills fungi.

Fungistatic. Action of a chemical that inhibits the germination of fungus spores while in contact.

Gallonage. Number of gallons of finished spray mix applied per acre, tree, hectare, square mile, or other unit.

General-use pesticide. A pesticide that can be purchased and used by the general public without undue hazard to the applicator and environment as long as the instructions on the label are followed carefully. (See *Restricted-use pesticide*).

Growth regulator. Organic substance effective in minute amounts for controlling or modifying (plant or insect) growth processes.

Harvest intervals. Period between last application of a pesticide to a crop and the harvest as permitted by law.

Hormone. A product of living cells that circulates in the animal or plant fluids and that produces a specific effect on cell activity remote from its point of origin.

182

Hydrolysis. Chemical process of (in this case) pesticide breakdown or decomposition involving a splitting of the molecule and addition of a water molecule.

Hyperplasia. Abnormal increase in the number of cells of a tissue.

Hypertrophy. Abnormal increase in the size of cells of a tissue.

Incompatible. Two or more materials that cannot be mixed or used together.

Ingest. To eat or swallow.

Ingredient statement. That portion of the label on a pesticide container that gives the name and amount of each active ingredient and the total amount of inert ingredients in the formulation.

Inhalation. Exposure of test animals either to vapor or dust for a predetermined time.

Inhalation toxicity. To be poisonous to man or animals when breathed into the lungs.

Insect-growth regulator (IGR). Chemical substance that disrupts the action of insect hormones controlling molting, maturity from pupal stage to adult, and others.

Insect pest management. The practical manipulation of insect (or mite) pest populations, using any or all control methods in a sound ecological manner.

Integrated control. The integration of the chemical and biological control methods.

Integrated pest management. A management system that uses all suitable techniques and methods in as compatible a manner as possible to maintain pest populations at levels below those causing economic injury.

Intramuscular. Injected into the muscle.

Intraperitoneal. Injected into the viscera but not into the organs.

Intravenous. Injected into the vein.

Invert emulsion. One in which the water is dispersed in oil rather than oil in water. Usually a thick mixture like salad dressing results.

Label. All printed material attached to or part of the pesticide container.

Labeling. Supplemental pesticide information that complements the information on the label but is not necessarily attached to or part of the container.

LC$_{50}$. The median lethal concentration, the concentration that kills 50 percent of the test organisms, expressed as milligrams (mg) or cubic centimeters (cc, if liquid) per animal. It is also the concentration expressed as parts per million (ppm) or parts per billion (ppb) in the environment (usually water) that kills 50 percent of the test organisms exposed.

LD$_{50}$. A lethal dose for 50 percent of the test organisms. The dose of toxicant producing 50 percent mortality in a population. A value used in presenting mammalian toxicity, usually oral toxicity, expressed as milligrams of toxicant per kilogram of body weight (mg/kg).

Leaching. The movement of a pesticide chemical or other substance downward through soil as a result of water movement.

Low-volume spray. Concentrate spray, applied to uniformly cover the crop, but not as a full coverage to the point of runoff.

mg/kg (milligrams per kilogram). Used to designate the amount of toxicant required per kilogram of body weight of test organism to produce a designated effect, usually the amount necessary to kill 50 percent of the test animals.

Microbial insecticide. A microorganism applied in the same way as conventional insecticides to control an existing pest population.

Mildew. Fungus growth on a surface.

Miscible liquids. Two or more liquids capable of being mixed in any proportions and of remaining mixed under normal conditions.

M.L.D. Median lethal dose (LD$_{50}$).

Molluscicide. A chemical used to kill or control snails and slugs.

Mutagen. Substance causing genes in an organism to mutate or change.

Mycoplasma. A microorganism intermediate in size between viruses and bacteria possessing many virus-like properties and not visible with a light microscope.

Necrosis. Death of tissue, plant or animal.

Nematicide. Chemical used to kill nematodes.

Oncogenic. The property to produce tumors (not necessarily cancerous) in tissues. (See *Carcinogenic*.)

Oral toxicity. Toxicity of a compound when given by mouth. Usually expressed as number of milligrams of chemical per kilogram of body weight of animal (white rat) when given orally in a single dose that kills 50 percent of the animals. The smaller the number, the greater the toxicity.

Organochlorine insecticide. One of the many chlorinated insecticides, e.g., DDT, dieldrin, chlordane, BHC, Lindane, etc.

Organophosphate. Class of insecticides (also one or two herbicides and fungicides) derived from phosphoric acid esters.

Ovicide. A chemical that destroys an organism's eggs.

Pathogen. Any disease-producing organism or virus.

Perennial. Plants that continue to live from year to year. Plants may be herbaceous or woody.

Persistence. The quality of an insecticide to persist as an effective residue due to its low volatility and chemical stability, e.g., certain organochlorine insecticides.

Pesticide. An "economic poison" defined in most state and federal laws as any substance used for controlling, preventing, destroying, repelling, or mitigating any pest. Includes fungicides, herbicides, insecticides, nematicides, rodenticides, desiccants, defoliants, plant growth regulators, etc.

Pheromones. Highly potent insect sex attractants produced by the insects. For some species, laboratory-synthesized pheromones have been developed for trapping purposes.

Physical selectivity. Refers to the use of broad-spectrum insecticides in such ways as to obtain selective action. This may be accomplished by timing, dosage, formulation, etc.

Physiological selectivity. Refers to insecticides that are inherently more toxic to some insects than to others.

Phytotoxic. Injurious to plants.

Piscicide. Chemical used to kill fish.

Poison. Any chemical or agent that can cause illness or death when eaten, absorbed through the skin, or inhaled by humans or animals.

Poison control center. Information sources for human poisoning cases, including pesticides, usually located at major hospitals.

Postemergence. After emergence of the specified weed or crop.

ppb. Parts per billion (parts in 10^9 parts) is the number of parts of toxicant per billion parts of the substance in question.

ppm. Parts per million (parts in 10^6 parts) is the number of parts of toxicant per million parts of the substance in question. They may include residues in soil, water, or whole animals.

Predacide. Chemicals used to poison predators.

Preplanting treatment. Made before the crop is planted.

Propellant. An inert ingredient in self-pressurized products that produces the force necessary to dispense the active ingredient from the container. (See *Aerosol.*)

Protectant. Fungicide applied to plant surface before pathogen attack to prevent penetration and subsequent infection.

Protective clothing. Clothing to be worn in pesticide-treated fields under certain conditions as required by federal law, e.g., reentry intervals.

Protopam chloride (2-pam). An antidote for certain organophosphate pesticide poisoning, but not for carbamate poisoning.

Rate. Refers to the amount of active ingredient material applied to a unit area regardless of percentage of chemical in the carrier (dilution).

Raw agricultural commodity. Any food in its raw and natural state, including fruits, vegetables, nuts, eggs, raw milk, and meats.

Reentry (intervals). Waiting interval required by federal law between application of certain hazardous pesticides to crops and the entrance of workers into those crops without protective clothing.

Registered pesticides. Pesticide products that have been approved by the Environmental Protection Agency for the uses listed on the label.

Repellent (insects). Substance used to repel ticks, chiggers, gnats, flies, mosquitoes, and fleas.

Residual. Having a continued killing effect over a period of time.

Residue. Trace of a pesticide and its metabolites remaining on and in a crop, soil, or water.

Resistance (insecticide). Natural or genetic ability of an organism to tolerate the poisonous effects of a toxicant.

Restricted-use pesticide. One of several pesticides, designated by the EPA, that can be applied only by certified applicators, because of their inherent toxicity or potential hazard to the environment.

RNA. Ribonucleic acid.

Rodenticide. Pesticide applied as a bait, dust, or fumigant to destroy or repel rodents and other animals, such as moles and rabbits.

Safener. Chemical that reduces the phytotoxicity of another chemical.

Secondary pest. A pest that usually does little if any damage but can become a serious pest under certain conditions, e.g., when insecticide applications destroy a given insect's predators and parasites.

Selective insecticide. One that kills selected insects, but spares many or most of the other organisms, including beneficial species, either through different toxic action or the manner in which insecticide is used.

Selective pesticide. One that, while killing the pest individuals, spares much or most of the other fauna or flora, including beneficial species, either through differential toxic action or through the manner in which the pesticide is used (formulation, dosage, timing, placement, etc.).

Senescence. Process or state of growing old.

Sex lure. Synthetic chemical that acts as the natural lure (pheromone) for one sex of an insect species.

Signal word. A required word that appears on every pesticide label to denote the relative toxicity of the product. The signal words are either *Danger— Poison* for highly toxic compounds, *Warning* for moderately toxic, or *Caution* for slightly toxic.

Slimicide. Chemical used to prevent slimy growth, as in wood-pulping processes for manufacture of paper and paperboard.

Slurry. Thin, watery mixture, such as liquid mud, cement, etc. Fungicides and some insecticides are applied to seeds as slurries to produce thick coating and reduce dustiness.

Soil application. Application of pesticide made primarily to soil surface rather than to vegetation.

Soil persistence. Length of time that a pesticide application on or in soil remains effective.

Soluble powder. A finely ground, solid material that will dissolve in water or some other liquid carrier.

Spot treatment. Application to localized or restricted areas, as differentiated from overall, broadcast, or complete coverage.

Spreader. Ingredient added to spray mixture to improve contact between pesticide and plant surface.

Sticker. Ingredient added to spray or dust to improve its adherence to plants.

Stomach poison. A pesticide that must be eaten by an insect or other animal in order to kill or control the animal.

Structural pests. Pests that attack and destroy buildings and other structures, clothing, stored food, and manufactured and processed goods; for example, termites, cockroaches, clothes moths, rats, and dry-rot fungi.

Stupefacient or soporific. Drug used as a pesticide to cause birds to enter a state of stupor so they can be captured and removed, or to frighten other birds away from the area.

Subcutaneous toxicity. The toxicity determined following its injection just below the skin.

Surfactant. Ingredient that aids or enhances the surface-modifying properties of a pesticide formulation (wetting agent, emulsifier, or spreader).

Suspension. Finely divided solid particles dispersed in a liquid.

Synergism. Increased activity resulting from the effect of one chemical on another.

Synthesize. Production of a compound by joining various elements or simpler compounds.

Systemic. Compound that is absorbed and translocated throughout the plant or animal.

Tank mix. Mixture of two or more pesticides in the spray tank at time of application. Such mixture must be cleared by EPA.

Target. The plants, animals, structures, areas, or pests to be treated with a pesticide application.

Temporary tolerance. A tolerance established on an agricultural commodity by EPA to permit a pesticide manufacturer or his agent time, usually one year, to collect additional residue data to support a petition for a permanent tolerance; in essence, an experimental tolerance. (See *Tolerance.*)

Teratogenic. Substance that causes physical birth defects in the offspring following exposure of the pregnant female.

Tolerance. Amount of pesticide residue permitted by federal regulation to remain on or in a crop. Expressed as parts per million (ppm).

Tolerant. Capable of withstanding effects.

Topical application. Treatment of a localized surface site such as a single leaf blade, on an insect, etc., as opposed to oral application.

Toxic. Poisonous to living organisms.

Toxicant. A poisonous substance such as the active ingredient in pesticide formulations that can injure or kill plants, animals, or microorganisms.

Toxin. A naturally occurring poison produced by plants, animals, or microorganisms; for example, the poison produced by the black widow spider, the venom produced by snakes, and the botulism toxin.

Trade name (trademark name, proprietary name, brand name). Name given a product by its manufacturer or formulator, distinguishing it as being produced or sold exclusively by that company.

Translocation. Transfer of food or other materials such as herbicides from one plant part to another.

Trivial name. Name in general or commonplace usage; for example, nicotine.

Ultralow volume (ULV). Sprays that are applied at 0.5 gallon or less per acre or sprays applied as the undiluted formulation.

Vector. An organism, as an insect, that transmits pathogens to plants or animals.

Virustatic. Prevents the multiplication of a virus.

Volatilize. To vaporize.

Weed. Plant growing where it is not desired.

Wettable powder. Pesticide formulation of toxicant mixed with inert dust and a wetting agent that mixes readily with water and forms a short-term suspension (requires tank agitation).

Wetting agent. Compound that causes spray solutions to contact plant surfaces more thoroughly.

Winter annual. Plant that starts germination in the fall, lives over winter, and completes its growth, including seed production, the following season.

Bibliography

Ashton, F. M., and A. S. Crafts. 1973. *Mode of Action of Herbicides*. John Wiley & Sons, New York. 504 pp.

Baily, J. B., and J. E. Swift. 1968. *Pesticide Information and Safety Manual*. University of California Agricultural Extension Service, Berkeley. 147 pp.

Beroza, M. (Ed.). 1970. *Chemicals Controlling Insect Behavior*. Academic Press, New York. 170 pp.

Borkovec, A. B. 1972. *Safe Handling of Insect Chemosterilants in Research and Field Use*. ARS, USDA (Agricultural Research Service, U.S. Department of Agriculture, Washington, D.C.). ARS-NE-2 (November).

Brazelton, R. W., N. B. Akesson, and W. E. Yates (Eds.). 1972. *The Safe Application of Agricultural Chemicals—Equipment and Calibration, Study Guide for Agricultural Pest Control Advisers*. Agricultural Publications, University of California, Berkeley. 58 pp.

Burges, H. D., and N. W. Hussey (Eds.). 1971. *Microbial Control of Insects and Mites*. Academic Press, New York. 861 pp.

Corbett, J. R. 1974. *The Biochemical Mode of Action of Pesticides*. Academic Press, New York. 330 pp.

Cummings, N. W. (Ed.). 1973. *Vertebrate Pests, Study Guide for Agricultural Pest Control Advisers*. Agricultural Publications, University of California, Berkeley. 125 pp.

Davies, J. E. 1976. *Pesticide Protection: A Training Manual for Health Personnel*. University of Miami School of Medicine, Miami, Florida. 71 pp.

Deal, A. S. (Ed.). 1972. *Insects, Mites and Other Invertebrates and Their Control in California, Study Guide for Agricultural Pest Control Advisers*. Agricultural Publications, University of California, Berkeley. 138 pp.

Degering, E. F. 1965. *Organic Chemistry*. College Outline Series. Barnes and Noble Books, New York. 422 pp.

Erwin, D. C. 1973. "Systemic Fungicides: Disease Control, Translocation, and Mode of Action." *Annual Review of Phytopathology* 11: 389–422.

Falcon, L. A. 1971. "Microbial Control as a Tool in Integrated Control Programs." Pp. 346–364 in *Biological Control*, C. B. Huffaker (Ed.). Plenum Press, New York.

Fowler, D. L., and J. N. Mahan. 1976. *The Pesticide Review 1975*. U.S. Department of Agriculture, Agricultural Stabilization and Conservation Service, Washington, D.C. 40 pp.

Hart, W. H. (Ed.). 1972. *Nematodes and Nematicides, Study Guide for Agricultural Pest Control Advisers*. Agricultural Publications, University of California, Berkeley, 53 pp.

Hays, W. J., Jr. 1975. *Toxicology of Pesticides*. Williams and Wilkins Co., Baltimore, Md. 580 pp.

Kenaga, E. E., and C. S. End. 1974. *Commercial and Experimental Organic Insecticides*. Special Pub. No. 74-1. Entomological Society of America, 4603 Calvert Road, College Park, MD 20740. 77 pp.

LaBrecque, G. C., and C. N. Smith (Eds.). 1968. *Principles of Insect Chemosterilization*. Appleton-Century-Crofts, New York. 354 pp.

Lukens, R. J. 1971. *Chemistry of Fungicidal Action*. Molecular Biology Biochemistry and Biophysics Series, No. 10. Springer Verlag, New York, 136 pp.

McCain, A. H. 1970. *Chemicals for Plant Disease Control (Fungicides, Nematicides, Bactericides)*. Agricultural Extension Service, University of California, Berkeley. 213 pp.

McHenry, W. B., and R. F. Norris (Eds.). 1972. *Weed Control, Study Guide for Agricultural Pest Control Advisers*. Agricultural Publications, University of California, Berkeley. 64 pp.

Mead, A. R. In press. "Economic Malacology with Particular Reference to *Achatina fulica*." *Annual Review of Entomology*.

Meister, R. T. (Ed.). 1978. *Farm Chemicals Handbook*. Meister Publishing Co., Willoughby, Ohio.

Melnikov, N. N. 1971. *Chemistry of Pesticides*. Springer Verlag, New York. 480 pp.

Menn, J. J., and M. Beroza (Eds.). 1972. *Insect Juvenile Hormones, Chemistry and Action*. Academic Press, New York. 341 pp.

Metcalfe, R. L., and J. J. McKelvey, Jr. 1976. *The Future for Insecticides.* John Wiley & Sons, New York. 524 pp.

Moller, W. J. (Ed.). 1972. *Plant Diseases, Study Guide for Agricultural Pest Control Advisers.* Agricultural Publications, University of California, Berkeley. 232 pp.

National Academy of Sciences. 1968. *Principles of Plant and Animal Pest Control. Vol. 1: Plant-Disease Development and Control.* Pub. No. 1596. Printing and Publishing Office, National Academy of Sciences, 2101 Constitution Ave., Washington, D.C. 20418. 205 pp.

National Academy of Sciences. 1968. *Principles of Plant and Animal Pest Control. Vol. 2: Weed Control.* Pub. No. 1597. Printing and Publishing Office, National Academy of Sciences, 2101 Constitution Ave., Washington, D.C. 20418. 471 pp.

National Academy of Sciences. 1969. *Principles of Plant and Animal Pest Control. Vol. 4: Control of Plant-Parasitic Nematodes.* Pub. No. 1696. Printing and Publishing Office, National Academy of Sciences, 2101 Constitution Ave., Washington, D.C. 20418. 172 pp.

O'Brien, R. D. 1967. *Insecticides: Action and Metabolism.* Academic Press, New York. 332 pp.

Olson, F. J. 1973. "The Screening of Candidate Molluscicides Against the Giant African Snail, *Achatina fulica* Bowdich." Master's thesis, Graduate College, University of Hawaii, Honolulu.

Orlob, G. B. 1973. "Ancient and Medieval Plant Pathology." *Pflanzenschutz-Nachrichten 26*(2). Farbenfabriken Bayer GMBH, Leverkusen, West Germany.

Pelczar, M. J., Jr., and R. D. Reid. 1972. *Microbiology.* 3rd ed. McGraw-Hill Book Co., New York. 948 pp.

Shorey, H. H. 1973. "Behavioral Responses to Insect Pheromones." *Annual Review of Entomology 18*:349–380.

Spencer, E. Y. 1973. *Guide to the Chemicals Used in Crop Protection.* Pub. No. 1093. 6th ed. Information Canada, Ottawa, Ontario, Canada. 452 pp.

Staal, G. B. 1975. "Insect Growth Regulators with Juvenile Hormone Activity." *Annual Review of Entomology 20:* 417–460.

Thomson, W. T. 1977. *Agricultural Chemicals. Vol. 2: Herbicides.* Thomson Publications, Indianapolis, Indiana. 264 pp.

Thomson, W. T. 1977. *Agricultural Chemicals. Vol. 3: Fumigants, Growth Regulators, Repellents, and Rodenticides.* Thomson Publications, Indianapolis, Indiana. 164 pp.

Torgeson, D. C., ed. 1967. *Fungicides, an Advanced Treatise. Vol. 1: Agricultural and Industrial Applications, Environmental Interactions.* Academic Press, New York. 697 pp.

Torgeson, E. Y. (Ed.). 1969. *Fungicides, an Advanced Treatise. Vol. 2: Chemistry and Physiology.* Academic Press, New York. 742 pp.

University of California. 1973. *Plant Growth Regulators, Study Guide for Agricultural Pest Control Advisers.* Agricultural Publications, University of California, Berkeley. 79 pp.

U.S. Environmental Protection Agency. 1975a. *Apply Pesticides Correctly: A Guide for Commercial Applicators.* Superintendent of Documents, U.S. Government Printing Office, Washington, D.C. 20402. 35 pp.

U.S. Environmental Protection Agency. 1975b. *EPA Compendium of Registered Pesticides: Vol. 1. Herbicides and Plant Regulators; 2. Fungicides and Nematicides; 3. Insecticides, Acaricides, Molluscicides and Anti-Fouling Compounds; 4. Rodenticides and Mammal, Bird and Fish Toxicants; 5. Disinfectants.* Technical Service Division, Office of Pesticide Programs, U.S. Environmental Protection Agency, Washington, D.C.

Van Valkenburg, W. 1973. *Pesticide Formulations.* Marcel Dekker, New York. 481 pp.

Ware, G. W. 1975. *Pesticides—An Auto-Tutorial Approach.* W. H. Freeman and Company, San Francisco. 205 pp.

Watson, T. F., L. Moore, and G. W. Ware. 1976. *Practical Insect Pest Management.* W. H. Freeman and Company, San Francisco. 209 pp.

Weaver, R. J. 1972. *Plant Growth Substances in Agriculture.* W. H. Freeman and Company, San Francisco. 594 pp.

Weed, Science Society of America. 1974. *Herbicide Handbook.* 3rd ed. Champaign, Illinois. 430 pp.

Westreich, G. A., and M. D. Lechtman. 1973. *Microbiology and Human Disease.* Glencoe Press, New York. 814 pp.

Wilkinson, C. F. 1976. *Insecticide Biochemistry and Physiology.* Plenum Press, New York. 768 pp.

Wiswesser, W. J. (Ed.). 1976. *Pesticide Index.* 5th ed. Entomological Society of America, College Park, MD 20470. 328 pp.

INVENTORY 1983